GENETIC
RECOMBINATION

Genetic Recombination

David R.F. Leach
Institute of Cell and Molecular Biology
University of Edinburgh

Blackwell Science

© 1996 by
Blackwell Science Ltd
Editorial Offices:
Osney Mead, Oxford OX2 0EL
25 John Street, London WC1N 2BL
23 Ainslie Place, Edinburgh EH3 6AJ
238 Main Street, Cambridge
   Massachusetts 02142, USA
54 University Street, Carlton
   Victoria 3053, Australia

Other Editorial Offices:
Arnette Blackwell SA
   1, rue de Lille, 75007 Paris
   France

Blackwell Wissenschafts-Verlag GmbH
   Kurfürstendamm 57
   10707 Berlin, Germany

   Feldgasse 13, A-1238 Wien
   Austria

First published 1996

Set by Excel Typesetters Company,
Hong Kong
Printed and bound in Great Britain
at The Alden Press,
Oxford and Northampton

DISTRIBUTORS

Marston Book Services Ltd
PO Box 87
Oxford OX2 0DT
(*Orders*: Tel: 01865 791155
        Fax: 01865 791927
        Telex: 837515)

USA
Blackwell Science, Inc.
238 Main Street
Cambridge, MA 02142
(*Orders*: Tel: 800 215-1000
        617 876-7000
        Fax: 617 492-5263)

Canada
Oxford University Press
70 Wynford Drive
Don Mills
Ontario M3C 1J9
(*Orders*: Tel: 416 441 2941)

Australia
Blackwell Science Pty Ltd
54 University Street
Carlton, Victoria 3053
(*Orders*: Tel: 03 9347-0300
        Fax: 03 9349-3016)

A catalogue record for this title
is available from the British Library

ISBN 0-632-03861-6

Library of Congress
Cataloging-in-Publication Data

Leach, David (David Reginald Francis)
   Genetic Recombination/David R.F.
   Leach.
      p.   cm.
   Includes bibliographical references
   and index.
   ISBN 0-632-03861-6
   1. Genetic recombination.   I   Title.
QH443.L43   1996
575.1'3—dc20   95-21294
                                  CIP

Contents

ERRATUM

Leach: *Genetic Recombination*

Fig. 4.3, p. 86 In the right hand panel of the figure, the value of Tw for the supercoiled DNA double-helix (the lower part of the figure) should be given as +4 not +3.

The supercoiled DNA should therefore have the values:

$Lk = +3$

$Tw = +4$

$Wr = -1$

In the legend to Fig 4.3 the equation should read:

$Lk = Tw + Wr$

Preface

In writing a book on genetic recombination my aim is twofold. Firstly, I want to introduce the concepts and experimental strategies that underpin this field, in a form understandable by anyone with a basic knowledge of genetics and biochemistry. Secondly I want to write a book including homologous, site-specific, transpositional and illegitimate recombination to demonstrate how the different aspects of the subject are intertwined. Furthermore, from a historical perspective, a wider understanding of recombination is timely now that transgenic technology has become so powerful and gene therapy of somatic cells promises new approaches in medicine.

In order to facilitate this, I explicitly discuss some experimental strategies. This is because our knowledge is based on such experiments and no real understanding can be complete without the ability to follow such experiments. However, to facilitate the assimilation of these experimental approaches, I have reduced the technical and system-specific details to a minimum. What is left are the experimental strategies themselves, and the reader is referred to the original papers to learn the details.

The classification of recombination reactions as homologous, site-specific, transpositional or illegitimate is the standard way of generating a logical subdivision of the field. I have therefore retained this system of classification, but I hope that the reader will develop an understanding that many recombination reactions utilize more than one of these mechanisms. For example, the transposition of Tn10 and P involves transpositional and homologous recombination; immunoglobulin VDJ-joining involves site-specific recognition and a reaction resembling illegitimate end-joining; mating-type interconversion involves site-specific cleavage and homologous recombination; and certain eukaryotic transposon-mediated genome rearrangements may be due to transposition coupled to illegitimate end-joining. I have used the term 'strand-transfer' in the sense that it is used by those working in the fields of transposition and site-specific recombination. I believe that this is novel when applied to homologous and illegitimate recombination, but creates a conceptual framework that is useful for describing all recombination reactions.

I have limited the use of references to specific experiments described in the figures and to figures adapted from other publications. There are no references in the main text but I provide suggestions for further reading at the end of each chapter. The ideas, interpretations and models presented here are all based on the significant contributions of others but, within the context of a book of this kind, I am unable to acknowledge all of these contributions individually. I apologize for this and refer the reader to the reviews listed. What I have done is to present my understanding of the field and this has involved some novel ways of representing recombination reactions.

I have been fortunate to have worked directly in three of the four areas of recombination described in this book. My PhD studies with Neville Symonds were on bacteriophage Mu transposition. I then worked on homologous recombination with Frank Stahl and finally I find myself investigating palindromic sequences that are unstable due to illegitimate recombination. Neville Symonds and Frank Stahl have undoubtedly contributed significantly to my understanding of this field. Neville Symonds has actively promoted an interdisciplinary approach to recombination by organizing, with Robin Holliday, the EMBO workshops on genetic recombination that were mostly held at Nethybridge in the Scottish highlands. These meetings have continued to be organized by Alain Nicholas and Steve West at Seillac in France and have contributed significantly to this book. Neville Symonds had a rule of no more than three slides per talk so that listeners could concentrate on understanding concepts, experimental design and conclusions instead of being presented with an indigestible amount of data. This book attempts to continue in this spirit.

I wish to thank Martin Boocock, David Brock, Raymond Devoret, Marty Gellert, Jim Haber, Hideo Ikeda, Roger Kemp, Ichizo Kobayashi, Steve Kowalczykowski, Noreen Murray, Charles Radding, Jim Shapiro, Dave Sherratt, Gerry Smith and Frank Stahl. Their advice, comments, criticism and support have made this venture possible. I have attempted to correct as many errors and omissions as possible but I take full responsibility for any that remain. The book is dedicated to the memory of my colleagues Ahmad Bukhari and John Scaife. I first met Ahmad on the train from Perth to Nethybridge in 1976 when I was a PhD student travelling to my first EMBO recombination workshop and was immediately captivated by his infectious excitement about Mu transposition. I became a colleague of John Scaife's when I joined the Molecular Biology department in Edinburgh in 1982. His wide range of interests never diminished the clarity of his vision and I had the pleasure of editing *Genetics of Bacteria* with him. Finally, I would like to thank Judith, Jonathan and Justine, for their love, support and forbearance throughout this project.

1 Genetic Recombination

1.1 Introduction

Genomes are not fixed in time and space; they combine the stability required for inheritance and the flexibility required for change. Genetic recombination is the rearrangement of genetic material. Systems exist to ensure that genomes are rearranged between generations and this results in genetic diversity between the individuals of a species. Genomes that are subdivided into several chromosomes have evolved, and this genomic arrangement facilitates the reassortment of genetic information. In sexually reproducing diploid organisms, homologous chromosomes segregate independently of each other at meiosis, enabling different combinations of genes to be present in the gametes. When male and female gametes fuse, an individual is formed that has a genome different from most if not all other members of the species. This happens by reassortment of chromosomes that are different from each other; a simple but important mechanism of recombination that does not require physical exchange (see Fig. 1.1).

The reassortment of genes on separate chromosomes does not require physical exchange but it does require the genetic diversity of chromosomes. If an individual's chromosomes were identical to those of the other members of its species, independent assortment and sexual reproduction would only produce genetically identical individuals. Two mechanisms underlie the genetic diversity of chromosomes. These are mutation and recombination involving physical exchange. This recombination can be subdivided into homologous, site-specific, transpositional and illegitimate mechanisms.

Genetic recombination does not only occur between generations. An individual's genome is also subject to rearrangement. This can result in the alteration of gene expression, the generation of genetic diversity between the cells in a single individual, or the repair of DNA damage. These events

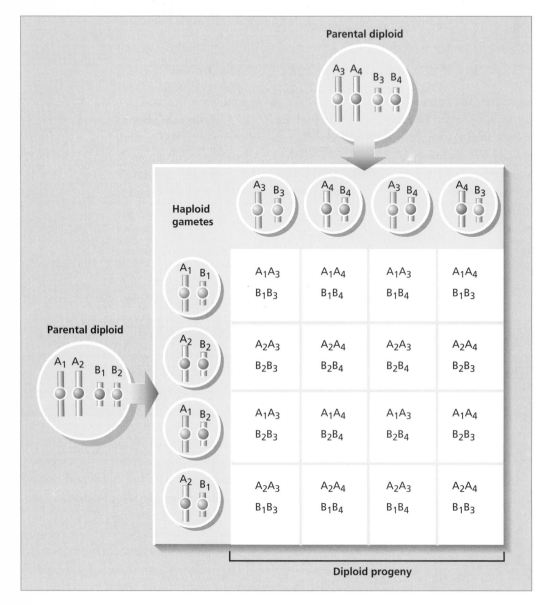

Fig. 1.1 Independent assortment of genes located on two chromosomes. This figure illustrates the variation that is generated by the independent assortment of chromosomes in a sexually reproducing diploid organism. Only two pairs of chromosomes (A and B) are represented. Maternal chromosomes (A_1, B_1, A_3 and B_3) and paternal chromosomes (A_2, B_2, A_4 and B_4) present in the parental diploids are shown in red and blue, respectively. Each parental diploid gives rise to four different haploid gametes and these fuse to produce 16 possible types of diploid progeny. As long as there are differences between the chromosomes A_1–B_4, the 16 progeny types will have different recombinant genomes. This reassortment is built into the reproductive system and does not require any physical exchange within chromosomes.

cannot occur by reassortment of chromosomes but do occur by the other mechanisms of recombination.

1.2 Homologous, site-specific, transpositional and illegitimate recombination

Recombination (by mechanisms other than reassortment) can be subdivided into four classes (see Fig. 1.2). **Homologous recombination** is defined by the use of DNA sequence homology for the recognition of recombining partners. Proteins (such as the RecA protein of *Escherichia coli*) may

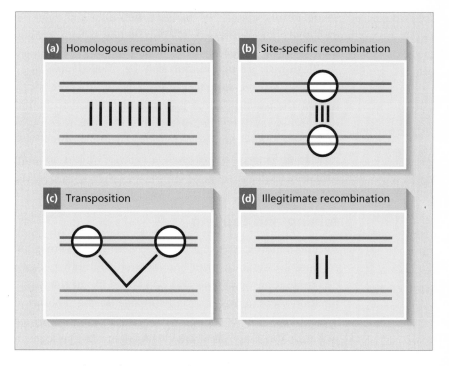

Fig. 1.2 Four classes of genetic recombination. (a) Homologous recombination involves the recognition of DNA homology between the recombining partners. The minimum length of homology required for homologous recognition differs for different organisms but is in the range of 25–300 bp. (b) Site-specific recombination is mediated by proteins that bind specific target sequences and catalyse recombination at those positions. The recognition of recombining sites involves protein–DNA and protein–protein interactions. (c) Transposition occurs for defined DNA sequences (transposable elements) that are recognized by transposon-encoded proteins responsible for recognition of the ends of the transposable element and target sequences. (d) Illegitimate recombination occurs at short DNA sequences, below the length required for homologous recombination. These sequences may share some homology but may also be non-homologous. Vertical and diagonal lines represent the recognition of the recombining partners and circles represent proteins recognizing specific DNA sequences.

facilitate this recognition but it is the base sequence that provides the specificity of recognition.

Site-specific recombination occurs when a protein binds a recognition site and catalyses exchange with another site that is also recognized by the same protein or another component of the same recombination system. There may be DNA homology between the recombining partners because they both bind the same protein or an intermediate step in the reaction requires homology. However, the basis of the recognition is not the DNA sequence; it is protein–DNA and protein–protein interactions that are crucial.

Transpositional recombination is similar in that it involves protein–DNA and protein–protein interactions but, here, only one of the recombining partners is specifically recognized by the protein that catalyses transposition. This protein recognizes the ends of the element accurately and then interacts with a relatively undefined target site.

Illegitimate recombination can be divided into two sub-classes: **end-joining** and **strand-slippage**. In both cases, however, the defining characteristic of the reaction is the presence of little or no homology between the recombining partners. Because of this, illegitimate recombination has sometimes been called non-homologous recombination. However, this term is misleading since there is often a role for micro-homology in illegitimate recombination.

1.3   Joining and copying

Recombination can take place by breaking and joining chromosomes or by copying information from one location to another. Often, in fact, there is a combination of joining and copying. One of the most primitive recombination mechanisms is strand-slippage during replication. This is a mechanism of illegitimate recombination and is discussed in Chapter 6. As shown in Fig. 1.3(a), a recombinant chromosome can be formed by replication of

---

Fig. 1.3 Strand-transfer in genetic recombination. (a) Illegitimate recombination by strand-slippage during replication. The newly replicated strand (shown in red) copies direct-repeat A, dissociates from its template and reassociates with direct repeat B. This recombination reaction, which has resulted in the deletion of the DNA between A and B has involved no joining of parental strands. There has been no strand-transfer. (b) Bacteriophage λ Int-mediated site-specific recombination. A pair of strands is broken, exchanged and religated to give the first pair of strand-transfers. These are shown by black dots where red and blue strands are joined. This first pair of strand-transfers generates a four-way junction known as the Holliday junction. Subsequently the strands that had not exchanged previously are cleaved, exchanged and religated in a second round of strand-transfer to give rise to the products of recombination. (c) Bacteriophage Mu transposition involves cleavage at the ends of the transposon (located within the blue DNA), staggered cleavage of the target

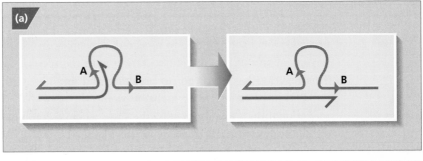

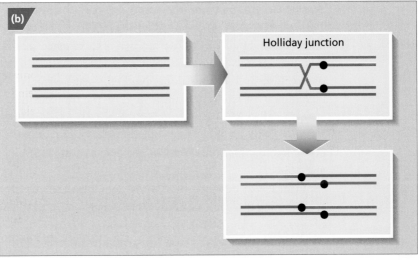

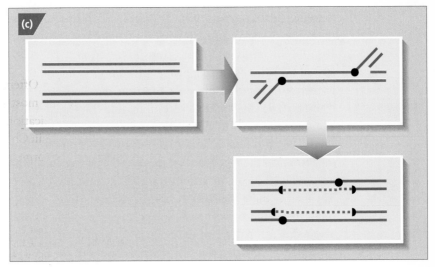

Fig. 1.3 (*Continued*) DNA (red) and joining of the transposon ends to the cleaved target. This joining reaction involves two strand-transfers marked by black dots and gives rise to two replication forks. DNA synthesis (dotted line) then completes the recombination reaction. This synthesis links the other two parental strands and therefore completes the second pair of strand-transfers. Since the region of newly synthesized DNA joins the two parental strands, these strand-transfers are represented by two half-dots at the ends of the newly replicated DNA.

one copy of a short, directly repeated sequence, dissociation of the newly replicated strand, and its reassociation with the second copy of the repeat. This reaction has involved no breaking and joining of strands, simply the replication of one part of a chromosome then another. A second type of recombination is exemplified by the site-specific recombination catalysed by the Int system of bacteriophage λ (see Fig.1.3(b)). Here, four DNA strands are broken and then re-joined in different combinations. The reaction passes through an intermediate where only one pair of strands has been exchanged. This is a four-way junction called the **Holliday junction**, a structure named after Robin Holliday who proposed its existence as an intermediate for homologous recombination. Recombination has been accomplished here, by breaking and joining alone. No DNA synthesis has taken place. A third recombination reaction is illustrated by the transposition of a transposable element called Mu (see Fig. 1.3(c)). Here, strands are broken and joined to a target. This creates two replication forks that allow DNA synthesis to occur. Recombination has taken place by a combination of breaking, joining and copying.

In order to describe the interactions between the DNA strands in the recombining molecules it is useful to define a term that describes the joining of a DNA strand of one recombining partner with that of another. Throughout this book the term **strand-transfer** has been used to describe this reaction. Please note that, in this sense, strand-transfer describes the **covalent** linkage of a DNA strand of one recombining partner to that of the other. We may now look at the recombination reactions illustrated in Fig. 1.3 in terms of strand-transfer reactions. The illegitimate recombination by strand-slippage has involved no joining of any DNA strands and is therefore an example of a reaction that has occurred without strand-transfer. The site-specific recombination catalysed by the Int system of bacteriophage λ is mediated by a pair of strand-transfer reactions to form the Holliday junction and a second pair of strand-transfers to resolve the junction into the recombinant products. The transpositional recombination has involved a pair of strand-transfers to join the transposon to its target. The intermediate then serves as a substrate for DNA synthesis and the newly replicated strands are finally joined to target. This second joining reaction also requires a pair of strand-transfers; the joining of newly replicated DNA to sequences flanking the new insertion point of the transposon. This second pair of strand-transfers almost goes unnoticed because it is simply the product of DNA polymerase and ligase activities rather than transposase itself. The concept of covalent strand-transfer is also useful in describing homologous recombination reactions (see Chapter 2).

Throughout this book, it is shown that all recombination reactions use the simple currencies of strand-transfer and copying reactions to mediate a wide variety of different events.

1.4

The soma and the germ-line

By the end of the 19th century, August Weissman had proposed that multicellular organisms were composed of two types of tissues: the **somatoplasm** and the **germplasm**. The somatoplasm (otherwise known as the **soma** or the **somatic tissue**) constitutes the functional components of the body while the germplasm (or **germ-line**) forms the gametes and is therefore directly responsible for the transmission of characteristics from generation to generation (see Fig. 1.4). This understanding was used to explain why acquired characteristics were not inherited, as had at one time been proposed and widely accepted.

Since the somatic tissue of animals does not contribute genetic material to progeny, the genome of somatic cells has some freedom to rearrange. It does so, for example in the lymphoid tissue, to generate the wide array of different immunoglobulin (Ig) molecules required for an effective immune response. There is clearly an advantage of this type of organization. A single organism can have both a relatively stable genome to pass on to its offspring and a variable genome to suit the needs of particular cell types. Such variability must, however, be well regulated. If it is not, it can result in cell death or uncontrolled (cancerous) cell growth. In plants, the situation is more complex. Whole individuals (including the germ-line) can be generated from certain somatic tissues and this must impose some restrictions on somatic recombination.

For most unicellular organisms their soma is their germ-line. In fact it is best to consider them as having only a germ-line. This is true whether the organism is exclusively haploid (e.g. the bacterium, *E. coli*) or has a diploid part of its life cycle (e.g. the yeast, *Saccharomyces cerevisiae*). The

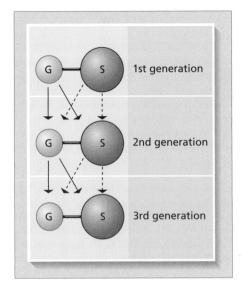

Fig. 1.4 **The germ-line and the soma.** The germ-line (represented by G) gives rise to the germ-line and soma (represented by S) of the next generation. In animals, the soma does not contribute information to the next generation. The consequence of this, is that genetic changes in the soma are not inherited. The somatic genome can be rearranged without affecting the chromosome organization passed on to the next generation. In plants, somatic tissue can give rise to both germ-line and soma, as shown by dotted lines.

word **most** is used above because at least one group of unicellular organisms has overcome this problem. This group is composed of the ciliates which are binucleate. One nucleus (the micro-nucleus) contains a genome which remains relatively constant and provides reproductive continuity and the other (the macro-nucleus) contains a genome that is extensively rearranged.

1.5   Meiosis

A simple grasp of meiosis is required to understand certain aspects of recombination. Sexually reproducing organisms must undergo meiosis to reduce the diploid to the haploid chromosome number. This permits the diploid state to be restored by fertilization. Meiosis is therefore essential for sexual reproduction. Meiosis proceeds via one round of DNA replication followed by two rounds of nuclear division. This is illustrated in Fig. 1.5. Initially, the chromosomes are replicated and by the leptotene stage this has been completed. Although replicated, the sister double-strands remain close to each other. By pachytene, the homologous chromosomes have come together and condensed to form **bivalents**. These bivalents contain the four double-stranded DNA molecules from the pair of replicated chromosomes but appear cytologically as a single entity. At diplotene, the bivalents start to come apart and can be seen to be composed of two homologous chromosomes, each of which has replicated to form two **chromatids**. At this stage, it is also possible to see points at which homologous recombination has occurred. These crossing points, or **chiasmata** (Fig. 1.6), are not recombination intermediates but are points at which crossing over has been completed and must not be confused with Holliday junctions. At

---

Fig. 1.5 Diagrammatic representation of meiosis. The essential features of meiosis are represented alongside micrographs of some the important stages. The micrographs are a compilation of different chromosomes. The line drawings have been made to illustrate the state of the DNA at the different stages and the positions of features such as the centromeres and crossovers at some stages are hypothetical. (a) Chromosomes are replicated by leptotene of the first meiotic division. The replicated copies lie together and each single blue or red line represents a DNA double-strand (i.e. there are four double-stranded DNAs for any one genetic locus. The centromere is represented by a circular dot. Under the microscope, the chromosomes appear as a tangled mess. During this early period, transient 'kissing' interactions occur between homologous chromatids. More stable contacts are then made and recombination is initiated at double-strand break sites (see Section 2.15). (b) The replicated homologous chromosomes pair throughout their lengths and by pachytene form tightly condensed bivalents. Each bivalent contains four double-stranded DNA molecules. At this stage recombination has occurred but this is not visible under the microscope. (c) At diplotene, the four chromatids become visible and points where crossing over is believed to have taken place can be seen as chiasmata. (d) The bivalents then orient on the metaphase plate and homologous centromeres are pulled towards the poles. (e) By anaphase, each

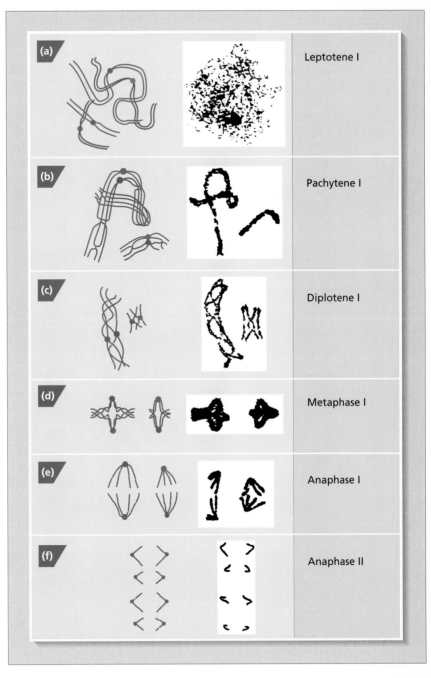

Leptotene I

Pachytene I

Diplotene I

Metaphase I

Anaphase I

Anaphase II

Fig. 1.5 (*Continued*) bivalent separates in two and chromosomes migrate towards the poles. A nuclear membrane then forms around the products of the first meiotic division. (f) The chromosomes then undergo a second division where sister centromeres separate to give rise to the haploid products. This division is similar to a mitotic division (but without replication) and is represented here at the anaphase stage where the chromosomes are pulled to the poles of the dividing nucleus.

metaphase 1, the bivalents lie in the equatorial plane and appear as crosses or circles as the centromeres start to pull the chromosomes apart. At this stage, homologous centromeres move apart but sister centromeres remain together.

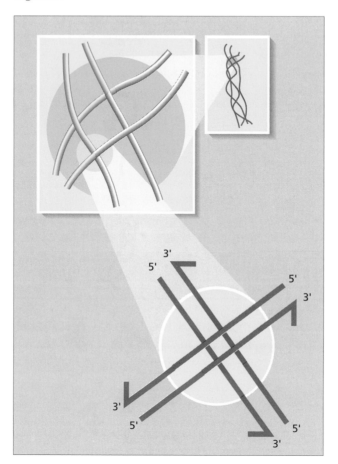

Fig. 1.6 Chiasmata as seen in bivalents at diplotene of the first meiotic division. By the time chiasmata become visible under the light microscope, recombination has been completed and they indicate the positions at which intact chromatids pass over each other. This figure illustrates the conventions used throughout this book to represent double- and single-strands of DNA. The first enlargement of part of the bivalent shows double-strands as cylinders crossing over each other. The second enlargement shows that each double-strand is composed of two single-strands. In reality, each single-strand is wound round its complement in a double-helix, but for simplicity they are represented as simple parallel lines. Each single-strand has a polarity that is shown running from a 5′ phosphate to a 3′ hydroxyl by a half-arrowhead at the 3′ end. In this figure, the joints between the two recombining chromatids are shown to have overlapping single-strands of heteroduplex DNA. The precise nature of these joints will depend on the mechanism of recombination, discussed in Chapter 2.

At anaphase 1, homologous chromosomes move to opposite poles and then a nuclear membrane forms around them. In the second meiotic division, the sister centromeres separate and the chromatids of each chromosome move to opposite poles. In this way, the haploid products are formed. Homologous recombination occurs at some point between leptotene and pachytene. It happens after the chromosomes have replicated and consist of four chromatids. Any one recombination event is therefore between two chromatids and the other two are not involved. The consequence of recombination between two linked genes is shown in Fig. 1.7. The genes are linked because they lie on the same chromosome and co-segregate unless separated by homologous recombination. Each of the genes is represented by a pair of **alleles** (*A* or *a*) and (*B* or *b*) that are also **markers**, since they mark genetic locations on the chromosome. It can be seen that recombination between these markers has resulted in two recombinant (*A,b* and *a,B*) and two parental (*A,B* and *a,b*) haploid products. Marker *A* has segregated from its allele *a* at the first meiotic division. This is because there has been no recombination between *A* and the centromere. On the other hand, marker *B* has not segregated from *b* at the first division. Instead, these markers have segregated at the second meiotic division. This is because recombination has occurred between *B* and the centromere. These segregation patterns are called **first-division segregation** and **second-division segregation**, respectively. The closer these genes lie together on the chromosome, the less frequently homologous recombination occurs between them and the closer they are said to be linked.

1.6   Recombination and biology

There must be a balance between genome flexibility and stability. The subdivision of the genome into chromosomes that can freely reassort during meiosis is a primary mechanism of ensuring this balance. This is a conservative mechanism. Linked genes remain together and unlinked genes are free to recombine at a high frequency. Homologous recombination between linked genes is also a relatively conservative mechanism. It retains linkage groups and closely linked markers recombine less frequently than distant markers. Furthermore, it can be used to repair chromosomes where information has been lost. Site-specific recombination can be more disruptive, since it is often involved in large-scale deletions or inversions. It is therefore restricted to specific situations where such changes are required. Transposition is very disruptive. Transposable elements can insert within genes and cause their inactivation. They can also mediate large-scale chromosomal rearrangements. Transposition is therefore tightly regulated to prevent the frequent occurrence of such events. Illegitimate recombination

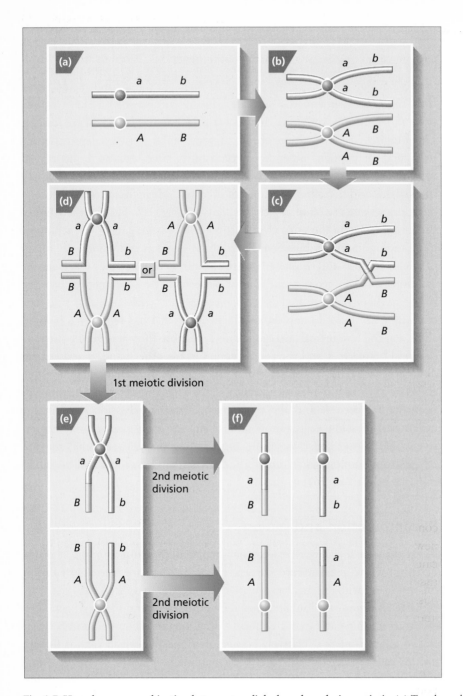

Fig. 1.7 Homologous recombination between two linked markers during meiosis. (a) Two homologous chromosomes carry markers represented by *A,a* and *B,b*. (b) Both chromosomes are replicated and form a bivalent composed of four homologous chromatids. (c) Recombination occurs between *A,a* and *B,b*. (d) Bivalents are oriented on the metaphase plate ready for the first meiotic division. Centromeres are ready to be pulled towards the poles. It can be seen that the bivalent contains two centromeres and there are two orientations in which these centromeres can lie with respect to the metaphase plate. These two orientations are equally likely and paternal and maternal chromosomes will orient independently of each other at this stage. Only one orientation is followed in (e) and (f). (e) The first meiotic division is completed and two recombinant chromosomes are formed. Markers *A* and *a* have segregated at this first division. (f) The second meiotic division is completed and four haploid products of meiosis are formed. Markers *B* and *b* have segregated at this second division.

is the most disruptive of all recombination reactions. It can cause deletions, amplifications, insertions and translocations with little regard for any base sequence requirements. It is therefore likely that illegitimate recombination is also tightly regulated. Nevertheless these most disruptive mechanisms of recombination are also the most able to produce rare events that can significantly affect evolution.

1.7   Recombination and genetics

Recombination is a cornerstone of genetics. The ability to map genes with respect to each other and then to use these maps to manipulate and understand the chromosomes of experimental organisms is at the heart of modern biology. Not only is homologous recombination widely used for such purposes but other recombination mechanisms are as well. Transposable elements are used to introduce genes into chromosomes or to inactivate genes and site-specific recombination systems are introduced into DNA sequences where specific recombination reactions are desired. Illegitimate recombination can be used to integrate any DNA sequence into a genome. The advent of **targeting** of eukaryotic genes by homologous recombination with introduced DNA sequences has opened up a new approach to the study of higher organisms. When coupled to the ability to regenerate a whole **transgenic** organism from a single cell containing a targeted gene, it is possible to ask questions about the action of specific genes in the context of the whole organism.

    Genetic mapping has allowed the study of genetic diseases and the identification of disease loci has permitted the **pre-natal diagnosis** of many conditions. Finally, recombination is contributing to the development of the new techniques of **gene therapy**, that promise to enable the treatment of the cause of genetic disease itself. The introduction of a wild-type gene into a tissue that is affected by a mutant gene promises to compensate for the missing or incorrect information. Many questions remain to be answered about such techniques and the methods themselves are likely to develop significantly over the next few years. Some questions that arise are: What is the likelihood of any adverse consequences of gene introduction by transposition or illegitimate recombination? Would re-introduction of homologously targeted cells be a preferable option? Are cells grown *in vitro* for homologous targeting equivalent to endogenous cells to be replaced. What are the moral, social and political consequences of these new technologies?

1.8   Conclusion

Mutation and recombination are the two processes that provide the starting material for evolution. Mutation causes genetic changes, that result in changes in the amino acid sequences of proteins, that result in phenotypic changes, that are acted upon by natural selection. Similarly, recombination provides the changes in the organization of the genome, that result in phenotypic changes, that can be acted upon by natural selection. Recombination reactions also play important roles in the specialization of somatic cells and in the repair of damaged DNA. This central position of recombination in biology is reflected in its central role in genetics. Mutation and recombination are the tools of the geneticist.

Further reading

Berg, D. E. & Howe, M. M. (eds) (1989). *Mobile DNA*. American Society for Microbiology, Washington.

Griffiths, A. J. F., Miller, J. H., Suzuki, D. T., Lewontin, R. C. & W. H. Gelbart (1993). *An Introduction to Genetic Analysis* (5th edn). Freeman, New York.

Kulcherlapati, R. & Smith, G. R. (eds) (1988). *Genetic Recombination*. American Society for Microbiology, Washington.

Moens, P. B. (ed) (1987). *Meiosis*. Academic Press Orlando, London.

Reference

John, B. & Lewis, K. R. (1973). The sexual system and the meiotic cycle. In: Head, J. J. (ed.) *The Meiotic Mechanism*. Oxford Biology Readers, Oxford University Press, Oxford.

2 Homologous Recombination

2.1 Introduction

Homologous recombination has been widely used to map linked genetic markers with respect to each other on chromosomes; that mapping works at all requires some degree of uniformity and predictability to the underlying reaction. On a gross scale, this is the case and the maps generated are of great value. However, when one starts to look at recombination on a minute scale, this uniformity and predictability breaks down. This is because recombination involves reactions that do not have precisely defined end-points and is catalysed by enzymes that have their own idiosyncratic specificities. These features of recombination have been most thoroughly investigated using bacteriophage λ, its host *Escherichia coli* and several fungi as model systems. These are the subject of this chapter.

Before the 1960s it was not known whether homologous recombination was accomplished by the replication, first, of part of one chromosome and then part of another, or whether recombination involved the breakage and reunion of chromosomes. These hypotheses have been called 'copy-choice' and 'break-join' and are illustrated in Fig. 2.1. The hybrid 'break-copy' hypothesis is also shown. When we ask ourselves the same question today we cannot give a unique answer. What we have learned is that there are many recombination systems that exist and that recombination intermediates can be processed in many different ways, depending on their structure and the enzymes available.

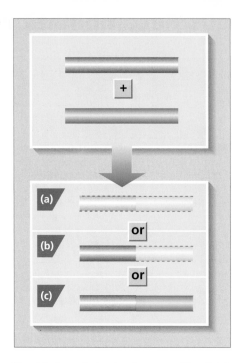

Fig. 2.1 Copy-choice, break-copy and break-join models of recombination. These hypotheses were initially suggested before it was known that DNA replicated semi-conservatively. In this figure, the essence of these three categories is presented without reference to whether the DNA is single- or double-stranded, how it is replicated, or how the joints are formed. Solid lines represent parental DNA and interrupted lines newly synthesized DNA. (a) Copy-choice recombination involves initially copying part of one chromosome and then the other. (b) Break-copy recombination implies the joining of a parental strand of DNA to a copy of the recombining partner. Cleavage of one parent must therefore occur and the broken end is used to prime DNA replication using the second parent as a template. (c) Break-join recombination requires no DNA replication and simply involves the cleavage and religation of parental strands.

We can sub-divide recombination reactions into three steps: initiation, repair and resolution. The first is **initiation**. There must be a mechanism by which base sequences are aligned in homologous register and, because recombination is precise, this must involve the recognition of one base sequence by another. Since DNA is normally double-stranded there are three possible interactions: single-strands with single-strands; single-strands with double-strands; and double-strands with double-strands as shown in Fig. 2.2. This may either result in the formation of heteroduplex DNA adjacent to one or more **Holliday junctions** (a and b) or if single-strand annealing occurs to no Holliday junction at all (d and e). If a Holliday junction is formed, the amount of heteroduplex DNA can then be increased or decreased by **branch-migration** (c). A number of studies have shown the

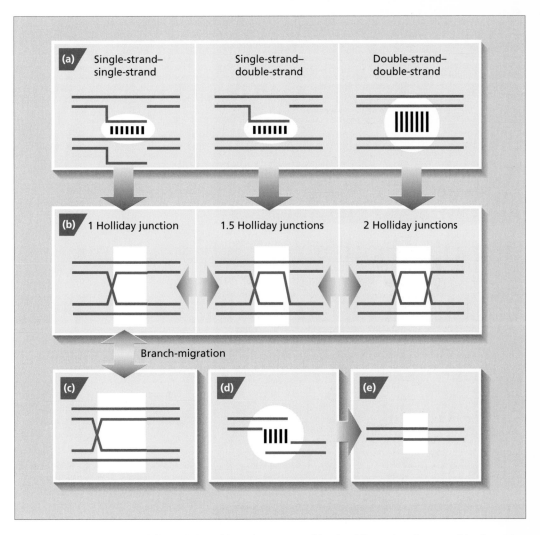

Fig. 2.2 The initiation of homologous recombination. We can imagine recombination to be initiated via the interaction of single-strands with single-strands, single-strands with double-strands or double-strands with double-strands, as shown in the regions circled in (a) and (d). The three representations are only examples of these classes of interaction (others would involve DNA with single-strand gaps, double-strand DNA ends, etc.). Initiation will lead to the formation of regions of heteroduplex DNA which are shown in the white boxes in (b), (c) and (e). The molecules drawn in (b) from left to right show one, one and a half, and two Holliday junctions. Branch migration to extend the region of heteroduplex DNA is shown in (c). Structures with more than a single Holliday junction can also branch migrate with the help of a topoisomerase (not shown). In (d) and (e) we can see that the annealing of single-strands can result in the formation of heteroduplex DNA and recombinant molecules without the formation of a Holliday junction.

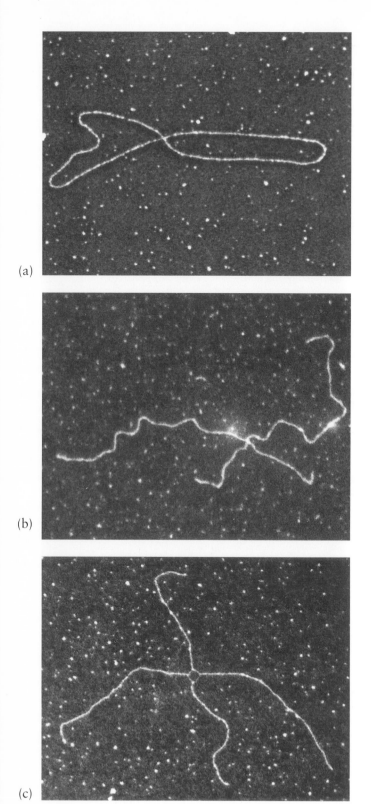

(a)

(b)

(c)

Fig. 2.3 Evidence for the existence of Holliday junctions in bacterial plasmids, (as in Potter & Dressler, 1977). Plasmid DNA was extracted from *E. coli* and shown to include a small proportion of Figure 8 molecules (a). This is the predicted conformation of two circular DNAs fused at the site of a Holliday junction. To prove that the DNAs were indeed joined via a Holliday junction, two further treatments were performed. The Figure 8s were digested by a restriction enzyme that cuts once in the plasmid, a treatment predicted to generate crosses with two pairs of arms of equal length if the junction was at a point of homology (b); and the DNA was spread under partially denaturing conditions expected to reveal the four strands present at the junction (c). Both treatments generated the predicted results. Micrographs kindly provided by H. Potter and D. Dressler (Harvard University).

existence of Holliday junctions in DNA isolated from cells. One example is shown in Fig. 2.3.

The second step is **repair of mismatched or missing information**. If the two recombining partners are not identical, mismatches will be generated within the heteroduplex DNA and these can be acted upon by mismatch-correction enzymes. Also, if information has been lost prior to, or during, initiation, this may be replaced by DNA synthesis (Fig. 2.4).

The third step, **resolution**, can be accomplished in a number of different ways depending on the structure of the intermediate and the enzymatic

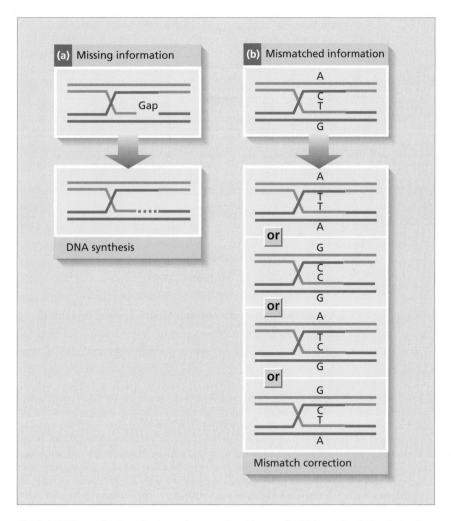

Fig. 2.4 DNA synthesis and mismatch correction. If part of a DNA molecule is missing it can be replaced by copying the information present on the recombining partner (a) and if mismatches are generated in the region of heteroduplex DNA these can be repaired by mismatch correction (b).

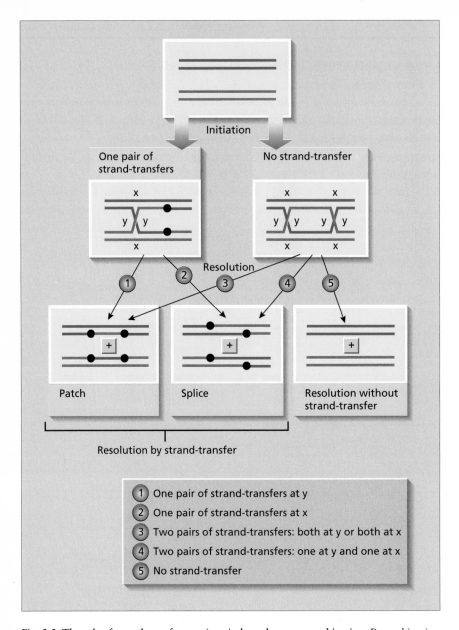

Fig. 2.5 The role of strand-transfer reactions in homologous recombination. Recombination can lead to the formation of two very different types of Holliday-junction-containing intermediates. The first is exemplified by a classical Holliday junction, where DNA strands from each of the parents are broken and rejoined to each other. The formation of this type of intermediate involves a pair of strand-transfers (shown as black dots where blue and red DNA strands are joined) and commits the molecules to an exchange of some kind. In order to resolve this structure, a second pair of strand-transfers is required either at the points marked 'x' or at the points marked 'y' to generate splice or patch recombinants respectively. The formation of either patch or splice recombinants require a total of two pairs of strand-transfers. The second type of intermediate is one with two (or an even number of Holliday junctions) and it can be formed without any exchange of phosphodiester backbones. This intermediate has the option of being resolved by two pairs of strand-transfers to generate patch or splice recombinants but can also be resolved by disentangling the DNA strands without exchange of backbones. This mechanism of resolution without strand-transfer would require catalysed branch-migration coupled to the action of a topoisomerase. Because it does not implicate any new DNA connections between the recombining molecules, the changes that do occur are restricted to the repair by DNA synthesis of any strands that are not complete when forming the intermediate and mismatch correction of any mismatches present in the heteroduplex DNA.

activities available. In Fig. 2.4, repair is depicted to occur prior to resolution, but this is not necessary. Repair can equally well take place on chromosomes that have already been resolved. The order of events will be determined by the availability of the necessary enzymes.

Firstly, let us take the situation where initiation has generated Holliday junctions. The initiation event will have led either to the formation of new phosphodiester backbone connections or not (see Fig. 2.5). These new connections are defined here as **strand-transfers**. See Chapter 1 (Section 1.3) for a more general introduction to the concept of strand-transfer as used here. If one pair of strand-transfers has occurred during initiation there will be an odd number of junctions, while if none have occurred there will be an even number of junctions. If the former is the case, resolution must occur by a second pair of strand-transfers (e.g. by cleavage of the junction followed by ligation to non-parental strands). If the latter has occurred then there are two possibilities for resolution. Either an even number of pairs of strand-transfers occurs, or no exchanges of phosphodiester backbone are made and the strands are separated by branch-migration. Whatever the mechanism of initiation, resolution by strand-transfer can result in the formation of both patch (non-crossover) and splice (crossover) recombinants depending on the strands that are cleaved and religated. By contrast, resolution without strand-transfer generates only recombinants that do not involve crossing over. Recombination is mediated exclusively by DNA synthesis, either as a consequence of mismatch repair or gap-filling (see Fig. 2.35, later).

The initiation events, shown in Fig. 2.2, all involved intact chromosomes with or without single-strand breaks. However, initiation can also involve chromosomes that have undergone double-strand cleavage and may have lost genetic information. In this situation, a Holliday junction may also be formed and its resolution leads to the formation of a replication fork as shown in Fig. 2.6. This can result in break-copy recombination.

Initiation events that do not create Holliday junctions, such as the annealing reaction shown in Fig. 2.2(d), will require different resolution steps to remove single-stranded ends that are likely to result from overlapping strands.

2.2   Recombination can involve chromosome breakage

Matthew Meselson, Jean Weigle and Grete Kellenberger set out to test whether bacteriophage λ recombinants could be formed by the breakage and reunion of chromosomes. The experiment of Meselson and Weigle (1961) is described in Fig. 2.7. This experiment demonstrated that at least some recombination occurred by breaking parental DNA strands. However,

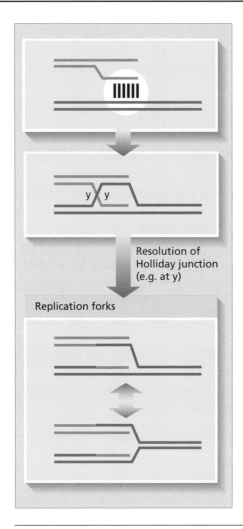

Resolution of
Holliday junction
(e.g. at y)

Replication forks

Fig. 2.6 Initiation of recombination by a broken chromosome and the formation of a replication fork by resolution of the Holliday junction. Initiation of recombination between a broken chromosome and its intact partner can lead to the formation of a Holliday junction. If this junction is resolved, this leaves a replication fork that can be used to replace the information that may have been lost from the broken chromosome.

Fig. 2.7 The Meselson–Weigle experiment (1968). This experiment concluded that recombination could occur via chromosome breakage by demonstrating the presence of parental atoms in recombinant progeny. Two λ phages were crossed in normal medium (medium containing the normal isotopes of carbon and nitrogen, ^{12}C and ^{14}N): one parent was wild-type (+,+), and was pre-labelled with heavy isotopes (^{13}C and ^{15}N); the other carried the mutations c and mi giving rise to clear, small plaques and was unlabelled. Because of the location of the genetic markers, the prediction of a recombination mechanism that involved chromosome breakage was that turbid, small plaque recombinants ($+$, mi) that had not undergone DNA replication would contain mainly ^{13}C and ^{15}N atoms, making them nearly as dense as the $+$, $+$ parent. Because replication was permitted during the cross, recombinants with hybrid and fully light density were also expected. They would not be informative concerning the mechanism of recombination since the incorporation of normal atoms could occur via a mechanism independent of recombination. In order to separate the progeny of the cross according to their density, the phage were centrifuged in an equilibrium gradient of CsCl. In the gradient, the phage particles band according to their density but do not lose their infectivity. Fractions from the gradient can therefore be titred for plaque-forming units. The results showed that there was a peak of $+$, mi recombinants that had a density almost as high as that of the $+$, $+$ parent. This demonstrated that they

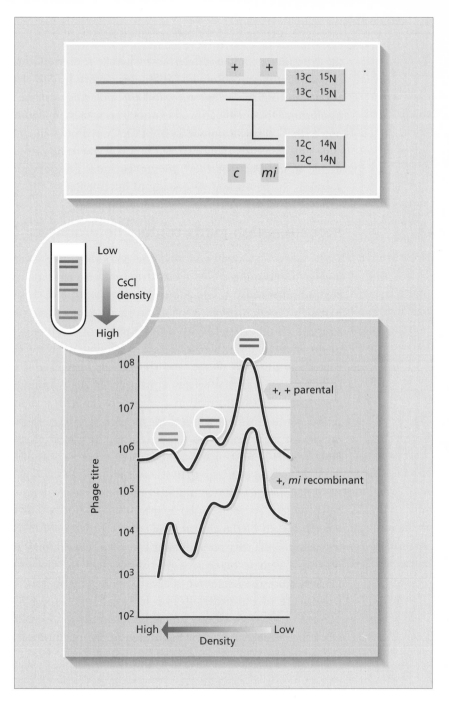

Fig. 2.7 (*Continued*) had inherited a majority of atoms from the two density labelled strands of the *+, +* parent. Since the cross was carried out in normal medium the small contribution from the *c, mi* parent could not be assessed for its origin via DNA replication or strand breakage. Therefore, the experiment did not distinguish between the break-join and break-copy models but did show that recombinants could be formed by a mechanism other than copy-choice. However, this conclusion must be tempered by the fact that it is restricted to the small proportion of recombinants that had not undergone DNA replication.

the experiment could not determine whether or not replication had occurred to complete the recombinant. Subsequent work by Meselson and by Frank Stahl has confirmed that recombinant molecules can be formed without significant DNA replication. However, even these experiments do not rule out a small contribution (< 3 kb) of DNA synthesis in the reaction. These experiments clearly demonstrate the involvement of chromosome breakage in recombination but they do not exclude a significant role for DNA synthesis in the normal processing of intermediates.

2.3   Recombination joints contain heteroduplex DNA

Early work had suggested that the recombination junction-point might involve overlapping DNA strands from the two parental molecules to form heteroduplex DNA. In phage crosses, this results in the potential for a single particle to carry different genetic information on each strand of its double-stranded chromosome. When a single phage of this type is plated on a lawn of sensitive bacteria, a plaque can result that is mixed (i.e. contains two genetically different phage) and this can be visualized if the heteroduplex includes a marker that affects a property of the plaque. For example in λ a $c/+$ heteroduplex (formed in a cross between a clear plaque-former and a turbid plaque-former) will appear mottled (i.e. have randomly distributed regions of clear and turbid appearance). In fungal crosses, heteroduplex DNA results in post-meiotic segregation of markers as described in Section 2.13. Meselson realized that if heteroduplex DNA was associated with recombination, there was a prediction about the structure of the recombinant products. As long as the length of heteroduplex was small, relative to the full length of the chromosome, then the amount of DNA contributed by each parent, to a recombinant carrying heteroduplex DNA at the position of a genetic marker, would depend on the position of the genetic marker. Since the λ genome is approximately 50 kb long and heteroduplex lengths are of the order of 1 kb in length, the prerequisite criteria were fulfilled in a system where the prediction could be tested and heteroduplex DNA was indeed found to be located at the position of crossing over, as shown in Fig. 2.8.

Fig. 2.8 **Evidence that heteroduplex DNA is formed at the site of recombinational exchange.** After Meselson (1965). Meselson performed a cross between density labelled (^{13}C, ^{15}N) wild-type phage and unlabelled phage carrying the clear plaque mutation c. As in the previous experiment, the cross was performed in normal medium (^{12}C^{14}N) under conditions permissive for DNA replication. The progeny were centrifuged in a CsCl equilibrium gradient, plated out and scored for plaque morphology. Clear and turbid plaques could be distinguished from plaques originating from heteroduplex DNA molecules at the c locus that form mottled plaques. If heteroduplex DNA were formed at the site of recombinational exchange, there would be a peak of mottled plaques at a density corresponding to the position of the c marker on the genetic map. Since c is located at $\frac{3}{4}$

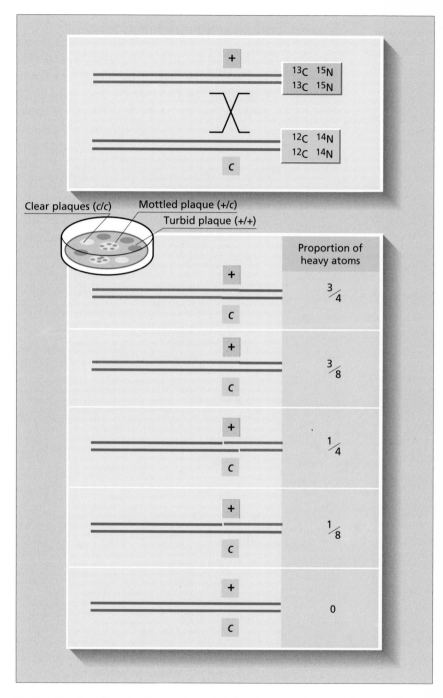

Fig. 2.8 (*Continued*) of the distance from the left end to the right end of λ we would expect mottled plaques with the following contributions of density labelled strands relative to the wild-type parent: $\frac{3}{4}$ and $\frac{1}{4}$ for crosses between unreplicated DNA, $\frac{3}{8}$ and $\frac{1}{8}$ for exchanges with wild-type DNA with one density labelled strand and unlabelled for exchanges with unlabelled wild-type DNA. The $\frac{3}{4}$ and $\frac{3}{8}$ labelled peaks are most easily resolved from parental material and were the peaks observed in this experiment.

Crosses and recombinant frequencies

Homologous recombination is usually studied by setting up crosses where two marked parental chromosomes are introduced into the same cell and allowed to recombine. The parental chromosomes are marked with mutations that affect the phenotype of the progeny in some way that can be

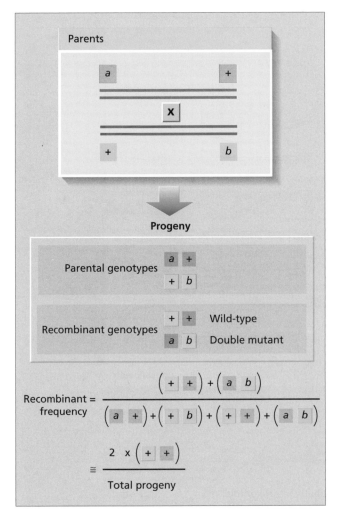

Fig. 2.9 The calculation of a recombinant frequency. The frequency of recombinants in a cross is defined as the yield of **recombinant progeny** as a proportion of the **total progeny**. In most crosses it is possible to select for a class of recombinants by virtue of their phenotype but there is often another class of recombinants that goes undetected. Since this is usually the case, recombinant frequencies are normally calculated by assuming that the two classes of recombinants will occur at an equal frequency in the population and that the frequency of +, + recombinants can be multiplied by two to obtain the recombinant frequency.

measured. The outcome of such a cross will be the formation of a particular fraction of recombinant chromosomes. The recombination frequency will be the frequency of recombinant chromosomes divided by the total number of chromosomes and this can be calculated by counting the number of progeny with particular genotypes. These recombinants will be of two types. They may, for example have combined two mutant genes and be double mutants or have combined two wild-type genes and have a wild-type genotype, as shown in Fig. 2.9. If so, it may not be possible to score the number of individuals carrying the double-mutant genotype because their phenotype is not different from the single mutants. For example, the single and double mutants may all be temperature sensitive. In this case, it is only possible to score the frequency of individuals with wild-type genotypes as a proportion of the total and we have to make the assumption that, in the population as a whole, the frequencies of double mutants and wild-types are the same. This is generally true unless the markers themselves affect the recombinant frequencies. Therefore, it is normally justified to multiply the wild-type frequency by two in order to estimate the recombinant frequency.

The cross described in Fig. 2.9 is known as a two-factor cross because two markers are used in the parental chromosomes between which recombination can be detected. However, crosses can be set up with more than two markers and these can be used to study the distribution of recombination events along the chromosome. A three-factor cross is shown in Fig. 2.10.

2.5   **Interference**

We can ask whether a recombination event appears either to stimulate or inhibit another event nearby; 'appears', because we shall see that two of the potential explanations for the clustering of exchanges between close markers do not imply the physical proximity of several independent recombination events. If recombination events were randomly distributed along chromosomes, and the occurrence of one event did not influence the probability of any other event in the same vicinity, the frequency of recombinants in an interval a–b multiplied by the frequency of recombinants in an interval b–c would accurately predict the frequency of double recombinants defined by these three markers, as shown in Fig. 2.10. If, however, there were less double recombinants than predicted from the frequencies of single recombinants, one could infer that the recombination events interfered with each other or showed **positive interference**. If, on the other hand, there were more double recombinants than predicted, it would appear that single events were clustered or showed **negative interference** (see Fig. 2.10). The

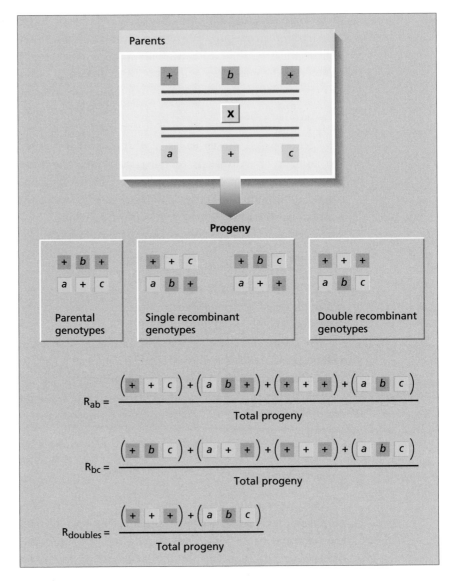

Fig. 2.10 Interference and the coefficient of coincidence. When performing a three-factor cross it is possible to determine whether the frequency of recombinants in each of the two intervals accurately predicts the frequency of double recombinants. If the recombinant frequency in interval a–b is R_{ab} and that in interval b–c is R_{bc} the predicted frequency of double recombinants (R_{doubles}) is $R_{ab} \times R_{bc}$

$$R_{\text{doubles}} = R_{ab} \times R_{bc} \quad \text{or:}$$

$$\frac{R_{\text{doubles}}}{R_{ab} \times R_{bc}} = 1 = S$$

This ratio is called the coefficient of coincidence and is denoted by the letter S. If $S < 1$ there are less double recombinants than predicted and recombination events are said to interfere with each other (positive interference). If $S > 1$ there are more double recombinants than expected and recombination events appear to be clustered (negative interference).

ratio of the observed to the expected frequency of double recombinants is a measure of the degree of association of recombination events. If this ratio (the **coefficient of coincidence** or S) equals 1, exchanges are random with respect to each other. If it is less than 1, there is positive interference. If it is greater than 1, there is negative interference. Interference itself can be measured by an index (I) where: $I = 1 - S$. I is positive for positive interference and negative for negative interference.

In eukaryotes, recombination between close markers (e.g. within a gene) shows negative interference but recombination between distant markers (e.g. in different genes) is generally characterized by positive interference. Paolo Amati and Matthew Meselson (1965) carried out bacteriophage λ crosses to determine the coefficient of coincidence as a function of the separation of the markers. They found that the coefficient of coincidence was always greater than 1 and increased as the markers used were closer together (Fig. 2.11). There are several possible explanations for the high level of negative interference that can be observed in crosses with close markers. Three of these are shown in Fig. 2.12.

1 Exchange events are indeed clustered (see Fig. 2.12a).

2 If the central marker (but not the flanking markers) is located within a region of heteroduplex DNA, this structure can be resolved to give rise to

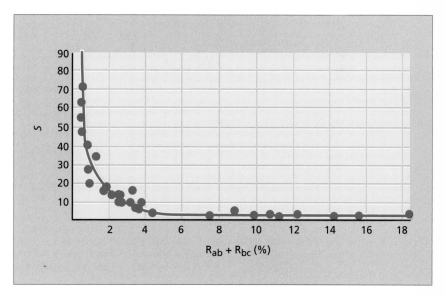

Fig. 2.11 The experiment of Amati and Meselson demonstrating the relationship between the coefficient of coincidence and the recombinational distance between markers. From Amati & Meselson (1965). Three factor crosses were performed between λ mutants and the coefficient of coincidence (S) plotted against the recombinational distance between the markers. As can be seen, the value of S increases dramatically as the markers are brought closer together.

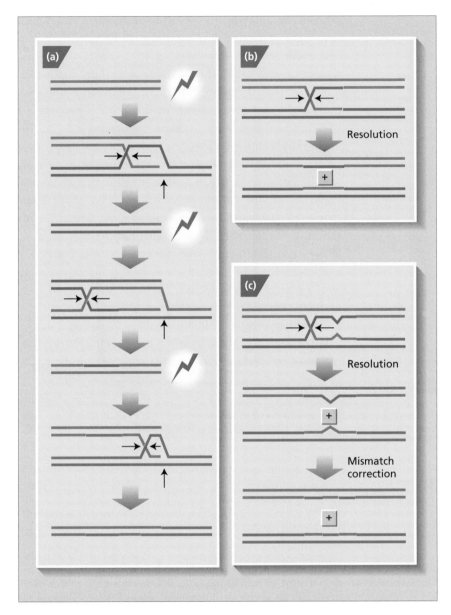

Fig. 2.12 Three explanations for the high negative interference observed with close markers. (a) **Clustering of recombination events.** The occurrence of a recombination event may stimulate the occurrence of recombination events nearby. This may be due to a local concentration of recombination enzymes or to the presence of a DNA structure that stimulates multiple rounds of recombination. An illustration of the latter situation is shown in this figure where one chromosome has a double-strand break and, at each round of recombination, one complete molecule and one broken molecule are generated by a mechanism such as that suggested in fig. 2.22. The broken products of the first two rounds of recombination are shown and these are expected to be highly recombinogenic. The complete product of the third round of recombination is shown containing a strand that

a double recombinant. This double recombinant has arisen not by the chance occurrence of two independent events but by the processing of a single intermediate. Therefore, it can be expected to occur at a frequency approaching that of the single event (see Fig. 2.12b). This mechanism can be used to explain an elevated frequency of close double recombinants but not triple- or higher-order recombinants.

3 If any of the markers lie within a stretch of heteroduplex DNA, they can be subject to mismatch-correction. If mismatch-correction occurs independently at each of the mismatches, multiple-exchange recombinants can arise from a single intermediate (see Fig. 2.12c).

In any particular organism, the contribution of each of these three mechanisms will depend on: the mechanism of recombination; the mechanism of mismatch-repair; and the length of time mismatches are accessible to repair, before the DNA is replicated.

2.6 Several recombination systems can co-exist within the same cell

At the time that Meselson was carrying out his experiments on recombination, it was not known that several recombination systems interact with phage λ DNA. Ethan Signer and Jon Weil demonstrated that a homologous and a site-specific recombination pathway are available to the wild-type phage as described in Fig. 2.13. The site-specific recombination pathway (Int pathway) is described in more detail in Chapter 4. The homologous recombination pathway (Red pathway) is catalysed by two phage-encoded products Redα and Redβ. Redα is a 5′ to 3′ exonuclease and Redβ is a protein that promotes the re-annealing of DNA single-strands. Together they act to promote recombination at the sites of double-strand breaks as described in Section 2.11. The *Escherichia coli* host cell also makes its own recombination proteins, but λ carries a gene, *gam*, responsible for interference with the host system. The host recombination genes also fall into at least two groups. One is only activated by a suppressor mutation and is located on a cryptic lambdoid prophage. This includes the genes *recE* and

---

Fig. 2.12 (*Continued*) once replicated would carry a triple exchange. Double-, triple- and higher-order exchanges are easily generated by this class of mechanism. (b) **The formation and resolution of a structure containing hybrid DNA creates strands that contain double exchanges.** Triple- and higher-order exchanges are not predicted as a consequence of this mechanism. (c) **Mismatch correction within a region of hybrid DNA.** In this case, as in (a), multiple exchanges are not restricted to double events. In (a), (b) and (c) the situations illustrated are only examples of how these three mechanisms can generate negative interference. The precise mechanism will depend on the rules for pairing, resolution and processing of junctions.

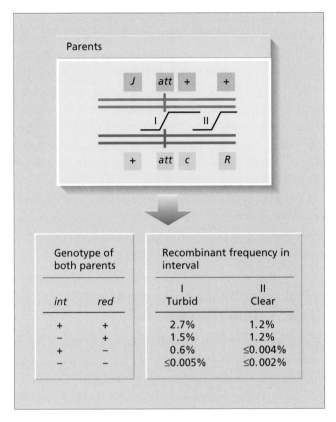

Fig. 2.13 Three recombination systems can operate on bacteriophage λ. Signer & Weil (1968 a & b) carried out crosses demonstrating that two recombination systems are encoded by bacteriophage λ. One, denoted Red, is a homologous system while the other, Int, is site-specific. Crosses were performed in *recA* hosts between *J*(*am*) and *c*, *R*(*am*) phage carrying mutations in either the *int* or *red* genes. In the crosses both parents had the same *int* or *red* genotype and the recombinant frequency in the interval between *J* and *c* and the interval between *c* and *R* were measured by counting the turbid and clear plaques respectively. It can be seen that *int* mutations modestly reduce recombination in the *J* to *c* interval but have no effect on the *c* to *R* interval. By contrast *red* mutations drastically reduce recombination in the *c* to *R* interval leaving a significant frequency of recombination in the *J* to *c* interval. *Int red* double mutants reduce recombination to background levels in both intervals. These results are explained by the action of the Int system at the *att* site of phage λ which is located between *J* and *c* while the Red system acts throughout the two intervals. Int mediated recombination is site-specific and Red mediated recombination is homologous. Since the crosses were performed in a *recA* host the *E. coli* recombination system was unable to operate. In *recA*⁺ cells the host system can operate on λ DNA but is partially inhibited by a phage encoded protein (Gam) which binds to and inactivates the RecBCD enzyme. Studies of the action of the host system on bacteriophage λ have therefore made use of *red gam* mutants.

recT which are analogous to the *red* genes of λ. The other is the primary set of *E. coli*'s recombination genes that are described below.

2.7   **The homologous recombination genes of *Escherichia coli***

In 1965, John Clark and Ann Margulies published the isolation of the first recombination defective mutants of *E. coli*. They mapped these mutations to a locus that has been called *recA* and is now known to encode a protein with homologous pairing properties—the RecA protein (see Section 2.8). Subsequently, other recombination defective mutants were discovered that identified the *recB* and *recC* genes. These genes encode two of the subunits of the RecBCD enzyme, exonuclease V (see Section 2.9). In addition, many other recombination genes have been identified that have overlapping functions. This means that they confer little or no recombination deficiency alone but do so in combinations. They include *recF*, *recO* and *recR* which may facilitate strand-pairing, recN which may protect recombinogenic ends in double-strand break repair, *recJ* which has 5′–3′ single-strand exonuclease activity and *recQ* which is a DNA helicase. Of particular interest are the gene *ruvC* that encodes a Holliday-junction-resolution protein and *ruvA*, *ruvB* and *recG* whose products catalyse branch-migration. Furthermore, the genes encoding DNA ligase, DNA polymerases, topoisomerase I, and single-strand-binding protein also contribute to homologous recombination.

2.8   The RecA protein

The biochemical activities of the RecA protein have been extensively studied, initially in the laboratories of Paul Howard-Flanders (1981) Robert Lehman, Tomoko and Hideyuki Ogawa, Jeff Roberts, Charles Radding, and subsequently in many others. A summary of its action in recombination is shown in Fig. 2.14. RecA is a protein of molecular weight 38 000 Da that polymerizes around single- and double-stranded DNA to form spiral **presynaptic filaments** (Fig. 2.15). The kinetics of binding to single-stranded DNA are more rapid than to double-stranded DNA and it is thought that single-strands are the primary signal for the formation of presynaptic filaments. The RecA in these filaments can, however, extend from single-stranded to double-stranded regions of the DNA. Presynaptic filaments have a polarity that is determined by the orientation of the RecA monomers that they comprise. A RecA monomer can be visualized as a molecule with a head and a tail, and it polymerizes in a head-to-tail chain along DNA, coating it in a 5′–3′ direction. Once formed, the filament has a polarity which is represented by the arrow in Fig. 2.14. The arrow is drawn in the orientation of RecA polymerization (5′–3′) but it is important to realize that it represents the intrinsic polarity of the filament, not just the polarity of its formation. Think of the RecA filament as a centipede composed of many polar segments.

When pre-synaptic filaments are mixed with naked double-strands they pair at regions of homology; a reaction described as **synapsis**. All three strands remain within the RecA filament and the DNA strands wind round each other in a three-stranded structure. This structure must be very

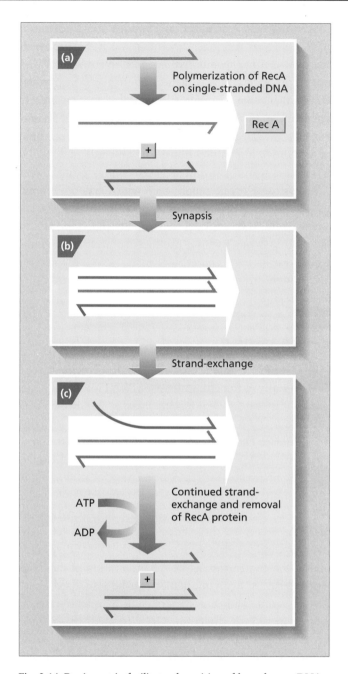

Fig. 2.14 RecA protein facilitates the pairing of homologous DNA sequences and the exchange of strands. The RecA filament is represented by an arrow and all helical turns of both protein and DNA have been removed to simplify the representation. (a) RecA protein forms a presynaptic spiral filament around single-stranded DNA and stretches it to 1.5 times its normal length. The presynaptic filament searches amongst uncoated double-stranded DNA for a homologous sequence. This search must involve the transient stretching of the double stranded DNA to 1.5 times its length and the testing of its sequence against

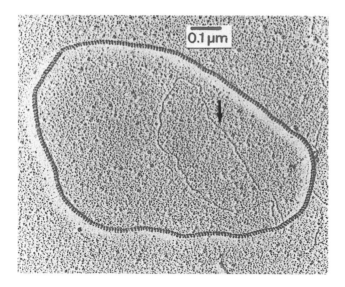

0.1 μm

Fig. 2.15 Electron micrograph of a spiral filament of RecA protein about a single-strand of DNA. The arrow indicates an uncoated circular single-strand. Micrograph kindly provided by A. Stasiak (University of Lausanne, Switzerland).

different from normal DNA since it is stretched 1.5-fold. The DNA strand that was originally within the presynaptic filament makes hydrogen-bonding connections with its complementary strand but the displaced strand remains closely associated to this newly formed duplex in a configuration that is insensitive to nuclease digestion. This may involve non-Watson–Crick base-pairing to the duplex. A provocative experiment, using oligonucleotides at the limit of the size normally required for homologous recombination, suggests that aspects of homologous recognition may be initiated via non-Watson–Crick hydrogen-bonding (Fig. 2.16).

After the strands have paired, a third reaction takes place where one strand of the DNA that was originally double-stranded is unpaired from the intermediate to generate the products: a duplex with one strand from each parent and a single-strand from the parental duplex. This reaction, which is polar with respect to the presynaptic filament, has been called **strand-exchange**. It does not require the hydrolysis of adenosine triphosphate (ATP) but is normally accompanied by ATP hydrolysis. The polarity

Fig. 2.14 (*Continued*) that of the presynaptic filament. Since the minimum length of homology required for RecA mediated recombination is of the order of 20 bp this may represent the length of DNA scanned for homology. (b) When a homologous sequence is found, a three-stranded intermediate is believed to form within the RecA filament. In this structure, hydrogen-bonding interactions are made between the strands that will be exchanged but all strands remain within the filament. (c) Strand-exchange can then occur. Strands that were originally paired emerge unpaired and strands that were originally unpaired emerge paired and ATP is hydrolysed to ADP. The hydrolysis of ATP causes the RecA protein to dissociate from the DNA.

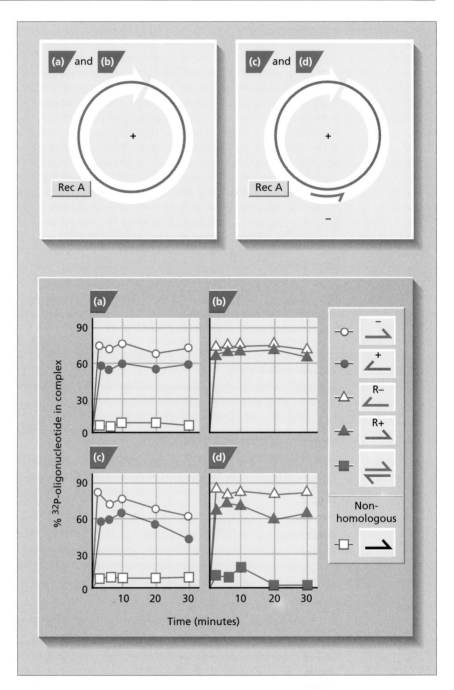

Fig. 2.16 Implication of non Watson–Crick bonding in synapsis. Rao & Radding (1993) performed an experiment in which a + strand of bacteriophage M13 was coated with RecA protein and this was then challenged with ^{32}P-labelled oligonucleotides that were either complementary, identical or non-homologous to 33 bp of the coated molecule and the amount of label retained in the complex was quantified. To their surprise, the identical sequence (+) was retained nearly as well as the complementary sequence (–) whereas the

of strand-exchange is defined with respect to the pre-synaptic filament. The reaction proceeds in the 5′–3′ direction with respect to the single-stranded region initially nucleating the binding of RecA protein (Fig. 2.17). Strand-exchange can proceed from a region involving three strands to one with four strands as shown in Fig. 2.18. This reaction is known as symmetrical strand-exchange.

2.9   *Chi* sites and the RecBCD enzyme

Frank Stahl and co-workers have shown that the distribution of RecA–RecBCD-mediated exchanges is uniform on λ DNA unless a *chi* site is present (Fig. 2.19). The *chi* site (also known as χ) causes a stimulation of recombination in its own vicinity that extends for a long distance (at least 10 kb). *Chi* is active when present on one or both parents (**dominance**), it stimulates recombination primarily to its left on the λ map (**direction-ality**) and it is only active in one orientation (**orientation-dependence**). The orientation-dependence of *chi* is determined with respect to the cohesive end site of λ (*cos*) which needs to be cleaved by the action of the enzyme terminase (Ter) for this *cos–chi* interaction to occur, as shown in the experiments described in Fig. 2.20. Gerry Smith and his colleagues sequenced several *chi* sites and showed that they all consisted of the octameric sequence 5′GCTGGTGG3′ reading left to right on the λ map.

The RecBCD enzyme (exonuclease V) is an ATP-dependent exonuclease and helicase. It binds to the ends of double-stranded DNA and unwinds the strands as it travels towards the centre. As the strands rewind behind it, loops are generated that increase in size as the enzyme travels through DNA (Fig. 2.21). The enzyme also cleaves DNA as it passes through it. The relative rates of unwinding and nucleolytic cleavage can be modulated by the concentrations of ATP, magnesium and calcium (higher ratios of ATP to magnesium reduce nuclease activity and calcium ions abolish nuclease activity). Gerry Smith, Andy Taylor and their colleagues have demonstrated that the cleavage reaction is also determined by the presence of the sequence 5′GCTGGTGG3′ (*chi*). A favoured position for nicking, under conditions

---

Fig. 2.16 (*Continued*) non-homologous sequence was retained poorly (see panel a). Next they investigated the effect of reversing the direction of the sugar-phosphate backbone. Here they found an even more surprising result. Both the 'identical' and the 'complementary' sequences with reversed backbones (R+ and R−) were retained as well as their normal counterparts (see panel b). In panels c and d, the same effect can be seen when double-stranded DNA is encased within the RecA presynaptic filament. These results imply that, at least in the cases where identical sequences are being recognized, this cannot be by conventional Watson–Crick base-pairing. They suggest that there is a mechanism that recognizes the sequence of bases independently of the polarity of the sugar-phosphate backbone.

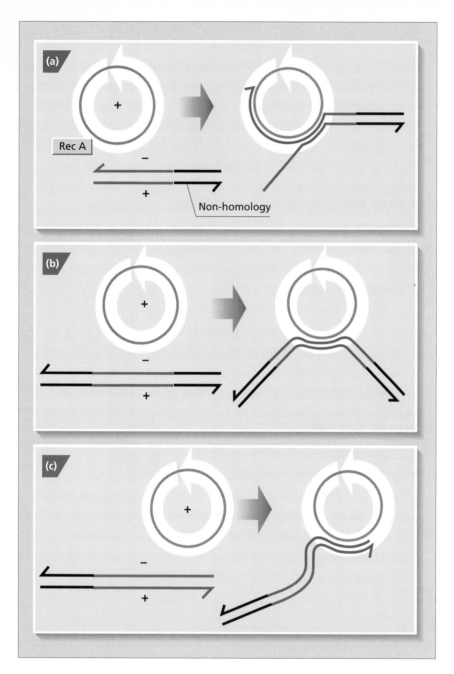

Fig. 2.17 Polarity of strand-exchange. Strand-exchange mediated by RecA protein proceeds with 5′–3′ polarity with respect to the RecA spiral filament. After Cox & Lehman (1981), Kahn, Cuningham, Das Gupta & Radding (1981) and West, Casuto & Howard-Flanders (1981)). Circular single-stranded DNA (+ strand) was incubated in the presence of RecA protein ATP and Mg^{2+} with homologous linear double-strands carrying blocks of non-homology on either one or both ends. Depending on the position of the non-homologous DNA, the reaction proceeded to different extents consistent with a 5′–3′ polarity of strand-exchange. This figure is a summary of the approach used and is based on several experiments. (a) When a homologous 3′ end (on the −strand) of the double-stranded DNA was available, synapsis and strand-exchange proceeded displacing the 5′ end. This reaction progressed 5′–3′ with respect to the RecA filament as defined by the polarity of the DNA in the presynaptic filament. The polarity of the RecA filament is represented by the large white arrow running 5′–3′. (b) When no homologous ends were present only synapsis was observed. (c) When a homologous 5′ end (on the −strand) was available synapsis was again observed but no strand-exchange could be detected.

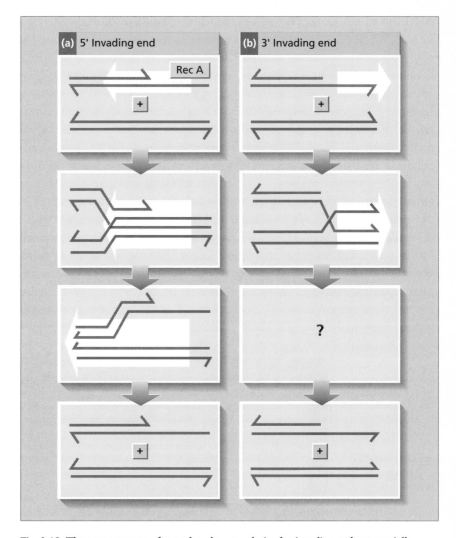

Fig. 2.18 The consequences of strand-exchange polarity for invading ends on partially single stranded DNA. If a single-strand with a 5′ end is available for RecA protein to polymerize onto, it will do so forming a filament pointing towards the region of double-stranded DNA. RecA will polymerize away from the end, leaving it uncoated. This configuration will favour strand-exchange to proceed into the region of double-stranded DNA (see a). By contrast, if a 3′ end is available, RecA will polymerize in the direction of the molecular end and coat it completely. The presynaptic filament will then pair with homologous double-stranded DNA to form a stable synaptic complex that may not initially favour strand-exchange (see b). However, because 3′ ends are well coated, they appear more reactive than 5′ ends and strand-exchange has been observed to proceed away from an invading 3′ end. How this occurs remains controversial since the initial presynaptic filament is inappropriately oriented for strand-exchange.

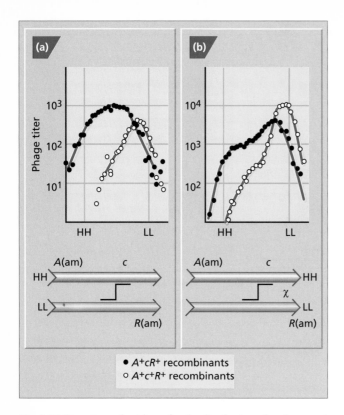

Fig. 2.19 Experimental evidence for the distribution of exchanges along the length of bacteriophage λ DNA when recombination occurs via the RecA–RecBCD system. Stahl & Stahl (1975) performed crosses between density-labelled *A(am)* phage and unlabelled *R(am)* phage under conditions where replication was blocked and centrifuged the progeny in a Cs-formate equilibrium gradient to separate them on the basis of their density. The *A* and *R* genes are located close to the ends of the λ chromosome. Therefore, in the absence of replication, the density of recombinants monitors the position of exchange events throughout the λ genome. The clear-plaque mutation *c* was included to enable the events occurring at the right end to be more clearly monitored. Both phages were *red, gam* and the host was *recABCD*⁺ to ensure that recombination proceeded via the Rec system. It can be seen in panel (a) that, in the absence of any *chi* sequence in either phage, the distribution of exchange events is approximately random. When *chi* is present in either one or both parents, as shown in (b), a stimulation of recombination in the vicinity of *chi* is observed. The *chi* site used in these experiments was located to the right of the *c* gene and stimulates recombination to its left. The expected migration position of phage particles, density labelled on both strands (HH), and unlabelled on both strands (LL), is shown on the graph.

Fig. 2.20 Experimental evidence for the basis of the orientation dependence of *Chi* (χ) in RecA–RecBCD mediated recombination. Kobayashi, Stahl and colleagues (1982, 1983 & 1985) showed that the orientation dependence of *chi* depended on the orientation of *cos* the cohesive end site of λ and the activity of terminase, the enzyme that cleaves at *cos* to initiate packaging into the phage head. To a first approximation, + indicates *chi* activity and − indicates no *chi* activity. (a) λ phage were constructed with an extra *cos* site cloned into the centre of the genome oriented in the opposite direction to that of the normal *cos*. These phage were then mutated at one of their *cos* sites and their interaction with *chi* observed. As shown in the figure, only oppositely oriented *cos* and *chi* were active. This suggested that some polar feature of λ biology in which *cos* was involved was implicated in the activity of *chi*. The two most obvious candidates were the polar packaging of λ that progresses from left end to the right end and the polar injection of λ that occurs from right to left. (b) Crosses were performed to test whether packaging of the recombinant from the *cos* site involved in the productive interaction with *chi* was required. These crosses again involved the use of phage with two *cos* sites, one to interact with *chi* and the other to ensure that the recombinant could be packaged even if the *cos* site interacting with *chi* did not have a partner required for completion of packaging. The *cos* site under investigation could either be present in *cis* with *chi* or in *trans*. As can be seen from the figure, oppositely oriented *cos* and *chi* can interact if they are located in *cis* but not in *trans*. However, the productive *cis* configuration does not require a partner-*cos* site and therefore could not have been used for packaging the recombinant. (c) Since packaging of the *chi*-stimulated recombinant was not required from the activating *cos* site, an experiment was performed to

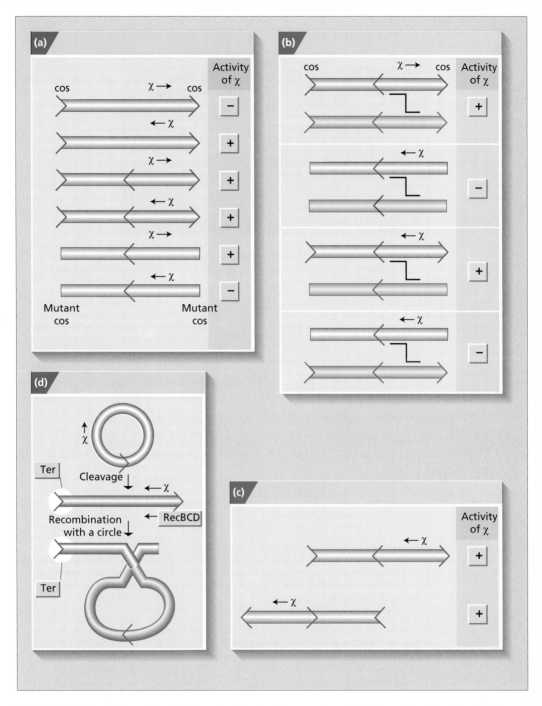

Fig. 2.20 (*Continued*) determine whether polar injection from *cos* determined the active orientation of *chi*. Phage lysates were prepared of a phage with two oppositely oriented *cos* sites but by genetic tricks were separated according to the *cos* site used for packaging. Each lysate therefore contained the same phage but packaged from a different *cos* site. These could be used to infect a new host to determine if *chi* was active. It was found that *chi* was active whichever *cos* site was used for injection of the DNA. The polarity of injection could therefore not be responsible for the orientation-dependent behaviour of *chi*. (d) These experiments established that the orientation-dependent activity of *chi* was not dependent on polar injection of λ DNA nor was it dependent on packaging from the *cos* site activating *chi*. A *cos* site in *cis* with *chi* was however required. Further experiments demonstrated that cleavage of the *cos* site in *cis* with *chi* was required for activation and it is believed that it is the generation of a double-strand break and the binding of terminase to one end of the break that are responsible for orientation dependence by determining the site of entry and direction of travel of the RecBCD enzyme.

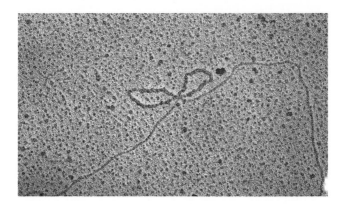

Fig. 2.21 RecBCD enzyme generates twin loops (rabbit ears) as it travels through DNA. As in Taylor (1988). Photomicrograph kindly provided by A. Taylor and G. Smith (Fred Hutchinson Cancer Research Institute, Seattle).

of reduced nuclease activity, is 4–6 nucleotides to the 3′ side of the *chi* sequence. Studies in Steve Kowalczykovski's laboratory have shown that, when more nuclease activity is allowed, cleavage of the DNA occurs frequently on the 3′ strand until the protein, travelling in the 3′–5′ direction with respect to the DNA strand containing *chi*, passes the *chi* sequence whereupon cleavage is reduced. Both laboratories have also shown cleavage of the strand opposite chi, at the position of *chi*, under conditions favouring nuclease activity. It is not yet clear which product of RecBCD action is the

Fig. 2.22 The nucleolytic activities of RecBCD enzyme on double-stranded DNA containing a *chi* site (χ) and their potential consequences for RecA binding. Under conditions of limited nuclease activity, RecBCD enzyme unwinds double-stranded DNA from an end and rewinds the strands behind itself. If a *chi* site is encountered, cleavage occurs 4–6 bp to the 3′ side of the *chi* octamer. Unwinding of the DNA strands then continues to give the product shown in (a). When more extensive nuclease activity is allowed, unwinding is accompanied by digestion, up to *chi*, of the DNA strand ending 3′ at the double-strand end attacked by RecBCD (b). When *chi* has been reached, further unwinding can generate the product shown in (c). Alternatively, if cleavage occurs opposite *chi* and unwinding continues, the product shown in (d) is formed. Each of the products (a–d) contains single-stranded DNA that can be used to initiate presynaptic filament formation by reaction with RecA protein. Three of the products (a, c and d) have two single-stranded regions of opposite polarity that have the potential to generate oppositely oriented presynaptic filaments. Whether one or both of these are important in recombination is unknown. The filament formed on the 3′ end must be built in a discontinuous mode, in the direction of RecBC(D) unwinding, because of the polarity of RecA protein polymerization. However, it will easily coat the DNA to the molecular end. It is this strand that forms the most stable joint-molecule with a double-stranded DNA in a coupled reaction containing RecBCD and RecA (Kowalczykowski *et al.*, 1994). The filament that can form on the single-strand with a 5′ end will grow continuously in the direction of RecBC(D) unwinding and has the

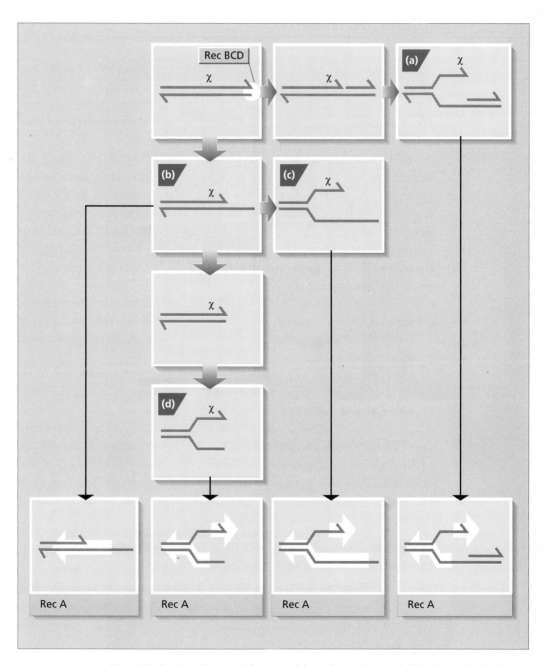

Fig. 2.22 (*Continued*) potential to spread from the single-stranded DNA into the double-stranded DNA. By contrast it can only grow discontinuously towards the molecular end which may remain uncoated. This filament is less reactive than that ending 3′ site of *chi* in the coupled RecBCD–RecA reaction when joint-molecule formation is assayed (Kowalczykowski *et al.*, 1994). However, the filament is oriented in the right direction to promote strand-exchange away from the site of joint-molecule formation. In this figure, RecBCD is shown entering a DNA molecule from the right end and moving towards the left. The *chi* site is oriented appropriately to be recognized by such a RecBCD molecule. The left end of the DNA could be infinitely long (e.g. part of a chromosome) and therefore not a site of RecBCD entry.

primary substrate for recombination. However, these results argue that RecBCD stimulates recombination by generating a substrate upon which RecA protein can act and that *chi* focuses the position of this activity. Four DNA structures generated by RecBCD action and their potential for interaction with RecA protein are shown in Fig. 2.22.

RecD mutants remain recombination proficient but the RecBC(D⁻) enzyme has lost all nuclease activity. However, it does retain some helicase activity. This suggests that the residual helicase activity of RecBC(D⁻) is important for recombination. Furthermore, density transfer experiments of the type shown in Figs 2.19 and 2.24 reveal that RecBC(D⁻)-mediated recombination is focused to the right end of λ. This has led to the proposal that *chi* reduces the nuclease activity of RecBCD enzyme by removing or otherwise inactivating the RecD subunit. When *chi* sites are made abundantly available *in vivo*, e.g. by simultaneous linearization of *chi*-containing plasmids, λ crosses become like crosses conducted in *recD* mutant cells. This mutant phenotype can be reversed by overproduction of the RecD protein in these cells.

Chi is an example of a recombination hot-spot or **recombinator**. It is a site specifically acted upon by a protein to stimulate recombination in its vicinity. Other recombinators are located at the sites of action of proteins whose primary function is not genetic recombination. An example of this type of interaction, described in Section 2.11, is the action of the enzymes of the Red system at double-strand breaks generated at *cos*. Sites that stimulate recombination in fungi have also been demonstrated and lead to the phenomenon of polarity described in section 2.14.

2.10

Enzymes that catalyse branch-migration and the resolution of Holliday junctions

Studies in the laboratories of Bob Lloyd, Hideo Shinagawa and Steve West have revealed the existence of *E. coli* enzymes that catalyse branch-migration and the resolution of Holliday junctions. The RuvC protein is a Holliday-junction-resolving enzyme. This protein has the ability to cleave Holliday junctions, as shown in Fig. 2.23, to generate recombinant products. Other enzymes with similar activities on Holliday junctions have been isolated from bacteriophages, yeast and mammalian cells.

Three *E. coli* proteins that promote b have been found. RuvA and RuvB work together to facilitate branch-migration in the direction of RecA-mediated strand exchange. RecG, on the other hand, primarily facilitates branch-migration in the opposite direction to RecA mediated strand exchange. Branch-migration is important for the formation of extended regions of heteroduplex DNA (see Fig. 2.2) and could be involved in resolution of intermediates via a mechanism such as that shown in Fig. 2.35. A model for Rec–Ruv–Pol-mediated recombination is shown in Fig. 2.23. This model proposes a hypothetical role for DNA synthesis which is clearly essential for the repair of double-strand gaps by recombination.

2.11   Double-strand breaks can act as recombinators

When Stahl and colleagues looked at the distribution of recombination when the Red system acts upon unreplicated λ, they found that it was not random. This situation was clearly different from that of Rec. In the case of Red, recombination was focused towards the ends of the λ chromosome (see Fig. 2.24). There were two possible explanations for this. Either, recombination was uniformly initiated along the chromosome but a recombination event located close to an end could more easily lead to a completed recombinant phage; or, the molecular ends of the chromosome in fact stimulated recombination. The first hypothesis can be understood in terms of a break-copy mechanism and the second by the action of the double-strand break at the end of the chromosome as a recombinator (Fig. 2.25). To distinguish between these two possibilities, the experiments shown in Fig. 2.26 were performed. It was concluded that the chromosome end (*cos*) could act to stimulate recombination in its own vicinity (i.e. it could act as a recombinator). In crosses where replication is permitted, exchanges are uniformly distributed along the chromosome and it is thought that this is due to the random position of double-strand breaks in replicating DNA.

2.12   Reciprocality of phage recombination

Reciprocal recombination is defined as follows. In a single recombination event between *AB* and *ab* parental chromosomes, recombination is reciprocal if both *Ab* and *aB* recombinants are generated. If only one of the possible recombinants (*Ab* or *aB*) is generated, recombination is non-reciprocal. At first sight this would seem easy to determine, however when looking at phage recombination there is a problem. We cannot look at the products of two individual chromosomes that have undergone recombination. This is because, in order to study recombination, we need to infect cells at a relatively high multiplicity to ensure co-infection with both parents; replication then occurs to generate many copies of each chromosome; encapsidation then picks out individuals from this pool; and finally, in a normal cross, we are looking at a population of recombinants arising from the infection of many cells. Therefore, if we see an equality of the yield of *Ab* and *aB* recombinants at the population level, this does not

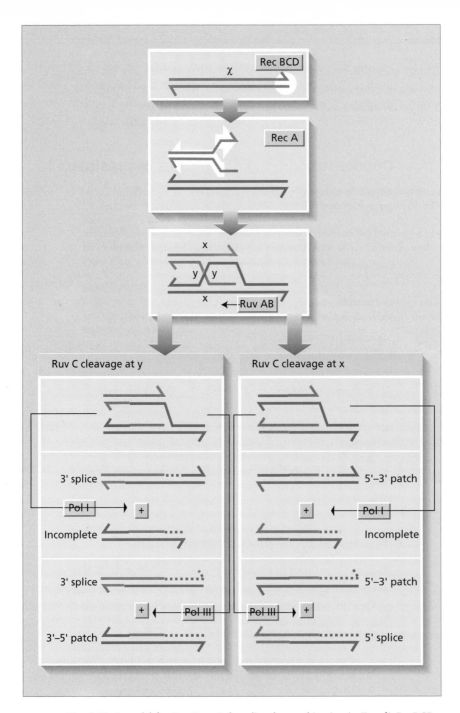

Fig. 2.23 A model for Rec–Ruv–Pol mediated recombination in *E. coli*. RecBCD enzyme enters DNA at the site of a double-strand break and travels through the molecule degrading it until it reaches a *chi* sequence (χ) whereupon its exonucleolytic activity is reduced and recombination is stimulated. RecA protein then forms one or more presynaptic filaments on the single-strand tails and one may extend into the double-stranded region. Homology is searched for and, when found, pairing occurs. RecA catalyses strand-exchange in the direction away from the broken end. This leaves an intermediate with one Holliday junction and a replication fork. The RuvAB proteins promote branch-migration of the Holliday

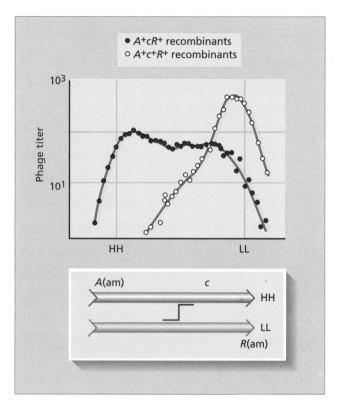

Fig. 2.24 Experimental evidence for the distribution of exchanges along the length of bacteriophage λ DNA when recombination occurs via the Red system. Stahl and colleagues (1974) performed replication blocked crosses between density labelled A(*am*) and unlabelled R(*am*) phage similar to those shown in Fig. 2.19. Here, however, the phage were *red*⁺ and the host Rec recombination system was inhibited by the phage Gam protein. In contrast to Rec recombination, the exchanges were not randomly distributed. Peaks of elevated recombination were observed close to the ends of the chromosome with the right hand end being more recombinogenic than the left. The expected migration positions of phage particles density-labelled on both strands (HH) and unlabelled on both strands (LL) are shown on the graph.

Fig. 2.23 (*Continued*) junction until a site of preferential cleavage for RuvC is found. Holliday-junction-resolution by RuvC leaves a replication fork and DNA synthesis proceeds using the free 3′ end as a primer. If synthesis is by DNA polymerase I, cleavage is likely to occur as soon as double-stranded DNA is reached by virtue of the 5′–3′ exonuclease activity of the protein. This results in the formation of a splice or a patch recombinant depending on the plane of cleavage by RuvC and an incomplete molecule that can undergo further rounds of recombination. If on the other hand DNA polymerase III is used, replication continues to the end of the chromosome to generate both splice and patch recombinants. If cleavage by RuvC occurs preferentially in the horizontal plane (y) rather than the vertical plane (x) there will be a bias in favour of 3′ overhangs in splice recombinants and 3′–5′ patch recombinants. Such a bias has been observed in bacteriophage λ crosses (Hagemann and Rosenbery, 1991 and Siddiqi *et al.*, 1991). This model is based on several previous models proposed by Steve Kowalczykowski, Susan Rosenberg, Gerry Smith and Frank Stahl but proposes specific but as yet hypothetical roles for DNA polymerases I and III.

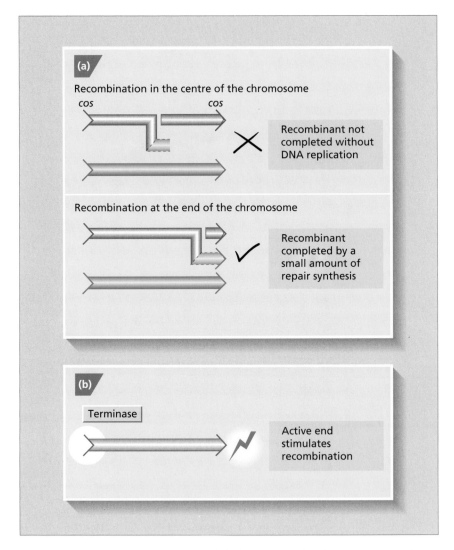

Fig. 2.25 Two potential explanations for the high frequency of exchanges near to *cos* in Red mediated recombination of unreplicated λ. After Stahl, Kobayashi & Stahl (1985). (a) Recombination can occur with equal probability throughout the λ genome. In the absence of extensive DNA replication, recombination events occurring near the ends can be more easily matured into packaged recombinants by limited DNA synthesis than those initiated in the centre of the chromosome. This explanation suggests that Red recombination proceeds via a break-copy mechanism. (b) The double-strand break generated at *cos* by the action of terminase is in itself recombinogenic. This explanation makes no prediction as to the involvement of replication in the event.

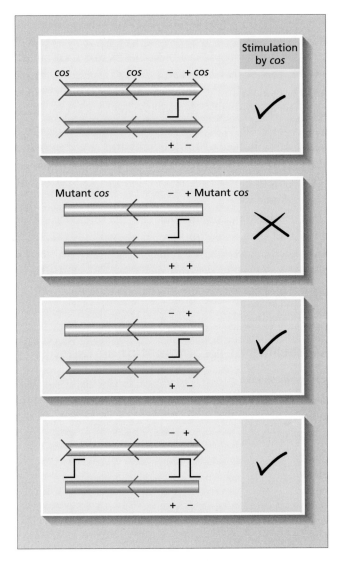

Fig. 2.26 The experiment performed to distinguish between explanations of the role of *cos* in Red mediated recombination of unreplicated λ. After Stahl, Kobayashi & Stahl (1985). A prediction of the break-copy model shown in Fig. 2.25(a) is that a second *cos* site must be reached by DNA synthesis in order for packaging to be completed. If this second *cos* site were deleted the stimulation of recombination observed at that end would be eliminated. In contrast, if *cos* acted as a recombinator, as shown in Fig. 2.25(b), the only requirement would be the presence of a single *cos* site to be cleaved by terminase. A second site for completion of the recombinant would not be necessary as long as an alternative route to packaging was provided. Such an alternative route was incorporated into the experiment by using phages with a cloned *cos* site in the centre of the chromosome. The results of crosses performed under replication-blocked conditions demonstrated that stimulation close to *cos* occurred whether or not a second site was present to complete the recombinant. Furthermore the vast majority of the recombinants that had undergone *cos*-stimulated recombination inherited the mutant *cos* when it was present. These results led to the conclusion that in Red-mediated recombination, *cos* acts as a recombinator due to its cleavage by terminase and that the information on the cleaved molecule is preferentially lost in the progeny.

necessarily imply reciprocality at the level of individual recombination events. What can be done, is to look at the products of individual infected cells (single bursts) and ask whether recombination is reciprocal there. However, these experiments still do not necessarily imply reciprocality or non-reciprocality at the level of individual recombining DNAs. This is because, even in a single cell, two non-reciprocal events can give the appearance of reciprocality; and a reciprocal event can give rise to non-reciprocal yields if one product of recombination is preferentially replicated and/or packaged. Although some work suggests that, at the level of single bursts, the Rec system may be reciprocal and the Red system non-reciprocal, the evidence is not strong.

Rec-mediated recombination in the presence of *chi* appears non-reciprocal at the population level as shown in Fig. 2.27, a result which can be explained by the enhanced nucleolytic activity of RecBCD prior to interaction with *chi*. However, this result may be caused by preferential packaging of the *chi*-free recombinant and argument still flourishes on this subject.

2.13   Reciprocality of recombination in fungi

The problems encountered in assessing the reciprocality of phage recombination highlight the power of fungal systems in the study of homologous

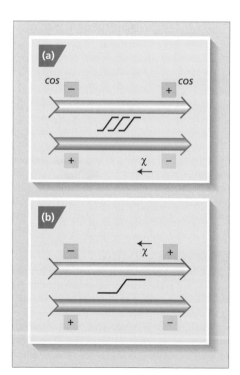

Fig. 2.27 Non-reciprocallity of *chi*-stimulated recombination, from Stahl, Stahl, Malone & Crasemann (1980). Replication-blocked crosses were performed in which the *chi* site stimulating recombination was either included in the +,+ recombinant (a) or not (b). The *chi*-stimulated recombinant lacking *chi* was recovered at a higher frequency than the *chi* containing recombinant. This is a non-reciprocal result at the level of the population of recombining molecules.

recombination. Here, the genotypes of the products of individual recombination events can be determined and this permits us to address the question of reciprocality directly. In the meioses of fungi, such as *Neurospora*, *Ascobolus* and *Sordaria*, each of the eight DNA strands involved gives rise to a spore that is positioned in an ascus in such a way that its origin can be deduced (Fig. 2.28). If a cross is made between a spore colour mutant and a wild-type strain, recombination in the vicinity of the spore colour locus results in different aberrant segregation patterns of the wild-type and mutant spores in the ascus. Some of these are shown diagrammatically in Fig. 2.29 where their origin is interpreted in terms of the formation of symmetric heteroduplex DNA followed by mismatch-correction. The eight spores in the ascus can be seen as four pairs of spores (a tetrad where each member of a pair corresponds to one strand of a DNA duplex). Therefore, if any pair is composed of non-identical individuals, there must have been an uncorrected mismatch which segregated at the post-meiotic division. This post-meiotic segregation is therefore evidence of heteroduplex DNA. From Fig. 2.30 it is clear that 6:2, 2:6, 5:3 and 3:5 segregation all represent non-reciprocal recombination. Aberrant 4:4 segregation by contrast remains reciprocal.

For close genetic markers non-reciprocal recombination is the rule whereas, for distant markers, recombination is usually reciprocal. Non-reciprocality for close markers often reflects the high probability of heteroduplex DNA extending across one or both of the markers and the

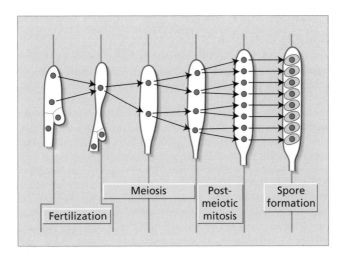

Fig. 2.28 Meiosis and spore formation in *Neurospora crassa*. (After Whitehouse, 1982.) After fertilization, *Neurospora* chromosomes undergo meiosis which is followed by a post-meiotic division to generate an ascus containing eight spores (an octad). These eight spores are arranged in four pairs (a tetrad of spore pairs) each of which is derived from one of the four chromosome copies (chromatids) in the bivalent undergoing meiosis.

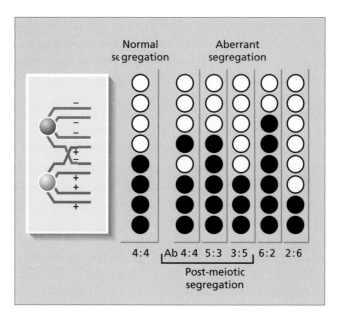

Fig. 2.29 Aberrant segregation patterns generated in a cross between a spore colour mutant and a wild-type fungal strain. In the absence of recombination, a cross between a wild-type strain and a spore colour mutant results in 4 : 4 segregation of wild-type and mutant spores (normal segregation). If, however, a recombination event has occurred close to the spore colour locus, a variety of aberrant segregation patterns can occur. In this figure, a symmetrical exchange of strands is represented which, if resolved at the Holliday junction, will result in aberrant 4 : 4 segregation. However, if one of the +/– mismatches is corrected, 5 : 3 (5 wild-type, 3 mutant) or 3 : 5 (3 wild-type, 5 mutant) segregation will occur. Alternatively, if both +/– mismatches are corrected, 6 : 2 (6 wild-type, 2 mutant) or 2 : 6 (2 wild-type, 6 mutant) segregation can occur. Asci showing 5 : 3, 3 : 5 and aberrant 4 : 4 segregation have spore pairs with non-identical members which must have arisen from segregation of wild-type and mutant information at the post-meiotic division. These asci are said to show post meiotic segregation and provide evidence for the existence of heteroduplex DNA at the locus in question. 6 : 2 and 2 : 6 asci retain no evidence of their formation by mismatch-correction of heteroduplex DNA and therefore may or may not have arisen via this route. Similarly 5 : 3 and 3 : 5 asci may not have arisen from symmetrical heteroduplex DNA in which case mismatch-correction is not required to explain their formation.

operation of mismatch-repair on any mismatches present. Non-reciprocal recombination can also result from physically non-reciprocal events such as those shown in Figs 2.33–2.35. If part of one chromosome is lost by nucleolytic degradation, non-reciprocality will result in genetic non-reciprocality for markers that have been lost. Since a recombination event may be reciprocal with respect to one pair of genetic markers and non-reciprocal with respect to another, reciprocality should always be defined with respect to the markers in question.

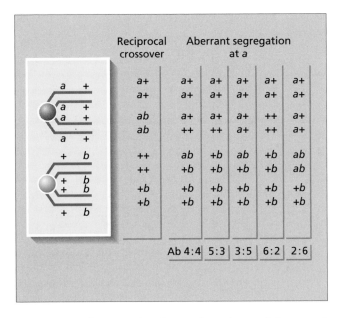

Fig. 2.30 Fungal crosses using close markers often result in non-reciprocal recombination. If we look at a cross between two strains carrying closely linked markers *a* and *b*, we can see that aberrant segregation at *a* leads to the formation of recombinant spores +, *b* and *a*, +. Aberrant 4:4 segregation leads to an equal yield of both types of recombinant spore, as does crossing over between the two loci. However 5:3, 3:5, 6:2 and 2:6 segregation at *a* yields unequal yields of +, *b* and *a*, + recombinants. This is non-reciprocal recombination. Many alternative routes (other that those shown in this figure) can also yield non-reciprocal recombination (e.g. normal segregation at *a* and aberrant segregation at *b*).

Historically, non-reciprocal recombination has also been described as **gene-conversion** or **conversion** a term which has sometimes been restricted to 6:2 and 2:6 segregation and at other times has included post-meiotic segregation. Terms such as half-chromatid conversion have also been applied to 5:3 and 3:5 segregation.

2.14 Polarity in fungal recombination

We saw that, in *E. coli*, *chi* sites stimulated recombination in their own vicinity and it is known that in fungi similar recombination hot-spots exist. Independently, Pascal Lissouba and Georges Rizet working with *Ascobolus immersus* and Noreen Murray working with *Neurospora crassa* showed that recombination frequencies varied predictably from one end of a gene to the other. There was a high-frequency end and a low-frequency end with

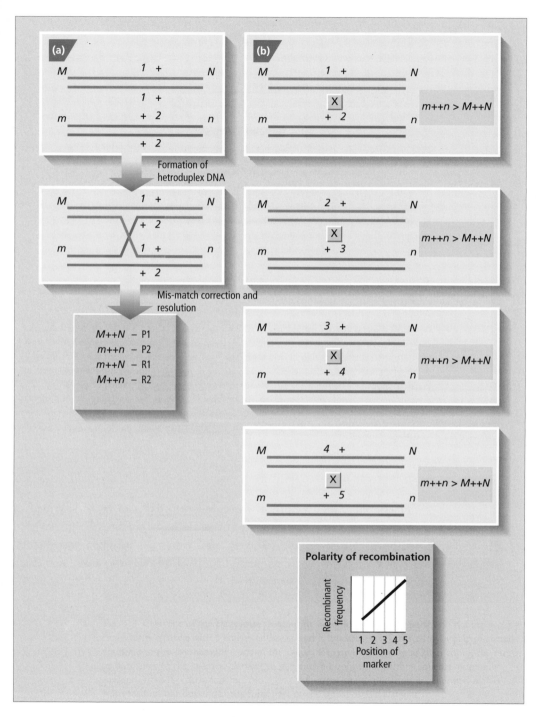

Fig. 2.31 The demonstration of polarity in *Neurospora crassa*. After Murray (1960 & 1963). Recombination can be studied in fungi by random spore analysis. When this is done, information is lost that would have been available had tetrad analysis been used. However,

a gradient in between. This is known as polarity. Polarity can either be studied by random spore analysis as shown in Fig. 2.31 or by tetrad/octad analysis as shown in Fig. 2.32. By using tetrad/octad analysis, it was possible to distinguish between the polarity of non-reciprocal recombination (6:2, 2:6, 5:3, 3:5 asci) and reciprocal recombination (aberrant 4:4 asci). Jean-Luc Rosignol and colleagues showed that, in *Ascobolus,* these two classes had different polarities (see Fig. 2.32). This is consistent with the Meselson–Radding or Aviemore model which is shown in Fig. 2.33. In this model, recombination is initiated asymmetrically to form asymmetric heteroduplex DNA which subsequently leads to symmetric heteroduplex DNA. All these genetic studies suggested that sites of initiation of recombination are located at the high-frequency ends of the genes.

2.15   Double-strand break repair

In yeast, several known recombination hot-spots correspond to sites of double-strand cleavage. These sites are found in regions of more open chromatin structure and have been shown in specific cases (e.g. *ARG4* and *HIS4*) to co-localize with the promoters of genes. It is also known that, in yeast, double-strand breaks stimulate recombination when linearized plasmids are introduced into the cell (see Chapter 7). Double-strand break repair (or **DSBR**) models have been proposed to account for this. Two examples of DSBR models that address the questions of resolution via

---

Fig. 2.31 (*Continued*) rare events can more easily be studied because recombinants can be selected for on specific media.

(a) In this cross, *M* & *N* are distant markers and *1* & *2* are close markers located within the same gene. Wild-type spores can be selected on a medium where the presence of the wild-type gene is required for growth. When this type of cross was performed an unexpected result was observed. The four classes of *++* recombinants (P1, P2, R1, R2) occurred at approximately equal frequencies. This result can be explained in terms of the formation of heteroduplex DNA, mismatch-correction and Holliday junction resolution as shown.

(b) Small differences in the frequencies of these classes do however occur. Noreen Murray looked at the frequency of P1 and P2 recombinants at the *me-2* locus of *Neurospora crassa* and observed that P2 > P1. There were two possible explanations of this. Either the nature or the position of the genetic markers affected the frequency of recombination. To distinguish these two hypotheses, crosses were made between pairs of mutants where the relative position of the mutations was known. In every case recombination was more frequent for the marker to the right-hand side of the gene (P2 > P1). This suggested that there was a polarity of recombination running from right to left across the gene and could be explained if recombination initiated on the right and heteroduplex DNA extended with decreasing probability towards the left. Markers on the right-hand side would, therefore, always have a higher probability of being included in heteroduplex DNA and consequently of recombination.

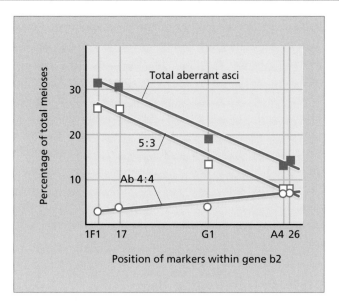

Fig. 2.32 Evidence for the Meselson–Radding model of recombination from the analysis of polarity in *Ascobolus immersus* using tetrad analysis. After Paquette N. & Rosignol J. L. (1978). Crosses were performed between wild-type *Ascobolus immersus* and mutants of the b2 locus and the progeny studied by tetrad analysis. Mutants that showed a high frequency of post-meiotic segregation were used since they provided evidence for the existence of heteroduplex DNA with little influence of mismatch correction. The figure shows that there is a polarity of recombination (total aberrant asci) that runs from left to right within b2. However, when the frequency of 5:3 and 3:5 asci was compared to that of aberrant 4:4 asci a significant difference was observed. The polarity of the former class followed that of the total polarity but the latter class if anything increased in frequency from left to right. Since mismatch correction had little effect on these crosses, it was concluded that the 5:3 and 3:5 asci derived from the formation of asymmetric heteroduplex DNA to the left of b2 and that this changed to symmetric heteroduplex DNA as it progressed to the right giving rise to aberrant 4:4 segregation.

Holliday-junction cleavage or branch-migration are shown in Figs 2.34 and 2.35. In these examples, 3′ invading ends are shown, but 5′ invading ends or blunt ends can be used to generate similar structures. In RecBCD-mediated recombination, the nature of the invading ends is not clear (see Fig. 2.22) whereas, in Red-mediated recombination and yeast recombination, evidence suggests that 3′ single-stranded ends are used. These models combine several features of the recombination reactions discussed in this chapter and illustrate how different outcomes may result from the processing of similar intermediate structures, depending on the nature of the enzymes present to process them. Key features that will determine the pathways of double-strand break repair are: the nature of the replication machinery; the branch-migration machinery; and the availability of the Holliday-junction-resolving enzymes.

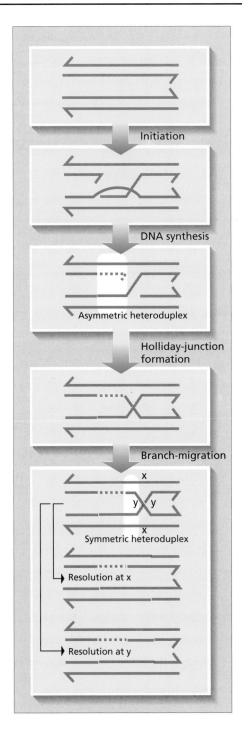

Initiation

DNA synthesis

Asymmetric heteroduplex

Holliday-junction formation

Branch-migration

x

y   y

x

Symmetric heteroduplex

Resolution at x

Resolution at y

Fig. 2.33 The Meselson–Radding model of genetic recombination. After Meselson & Radding (1975). Recombination is initiated with an asymmetric invasion by a DNA strand from one parent. This displaces one of the resident strands which is then degraded. The result is a gap on the chromosome which initiated recombination and this is repaired by DNA synthesis. Asymmetric heteroduplex DNA has thereby been created which, if uncorrected, will lead to 5:3 or 3:5 segregation. If at this point a Holliday junction forms and migrates away from the initiation point, symmetric heteroduplex DNA will result giving rise to aberrant 4:4 segregation in the absence of mismatch correction. Finally the Holliday junction is resolved in the two possible modes to give rise to patch and splice recombinants. The splice recombinants account for crossing over of distant markers while the patch recombinant account for the high frequency of close marker exchange that occurs in the absence of crossing over.

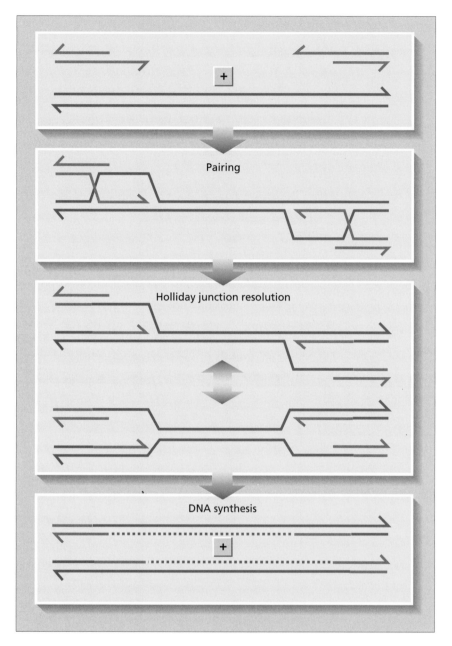

Fig. 2.34 Double-strand break repair (DSBR) by Holliday junction resolution. After Szostak, Orr-Weaver, Rothstein & Stahl (1983). In this model, recombination is initiated at the site of a double-strand break. The double-strand break is then attacked by nucleases to leave frayed ends. These invade a homologous duplex and DNA synthesis occurs using the 3′ end as a primer. This repair synthesis results in the formation of two Holliday junctions adjacent to two replication forks. The junctions can be resolved by cleavage in either of the two possible orientations. Depending on the orientation of Holliday-junction cleavage, replication will result in the formation of splice or patch recombinants. Only the splice

2.16   ## Conclusion

We have seen that double-strand breaks are very important in the two best-studied prokaryotic systems, Red and Rec. In Red-mediated recombination double-strand breaks act to stimulate recombination in their own vicinity whereas, in Rec-mediated recombination, double-strand breaks permit the entry of the RecBCD enzyme that stimulates recombination at and beyond the position of *chi*. Double-strand breaks are also recombinogenic in yeast. It does makes sense that double-strand breaks should stimulate recombination since double-strand breaks are easily converted to double-strand gaps and recombination is the only way of ensuring the accurate repair of this type of DNA damage. Indeed, if formed without strand-transfer and resolved by branch-migration, as shown in Fig. 2.35, repair could be accomplished without the danger of crossing over. Danger, because crossing over between homologous sequences at different chromosomal locations will cause gross DNA rearrangements such as translocations, deletions and inversions. We could therefore speculate that Holliday-junction-resolving enzymes and recombination via strand-transfer might be tightly regulated. The answers to this and many other questions await further investigation. Now that homologous recombination can be detected and studied in mammalian cells, we can look forward to understanding the mechanisms involved. However, if there is one lesson to be learned from the better understood microbes, it is that several enzymatic pathways can co-exist within a single cell.

Fig. 2.34 (*Continued*) product is shown here. This model provides a route to the formation of 6:2 and 2:6 asci without invoking mismatch correction of hybrid DNA. One member of a pair of chromatids, present in a bivalent when recombination occurs, acquires (on both of its strands) the characteristics of the other pair. We are therefore left with three chromatids of one type and one of the other (a 3:1 or 6:2 configuration). Remember when looking at the figure that recombination in meiosis occurs when chromosomes have duplicated and that there are two chromatids that are not participating in the recombination event and have not been depicted. Evidence from yeast recombination suggests that 3:1 tetrads generally arise from mismatch-correction. This supports the argument that, in yeast, the degradation of 3' ends is limited.

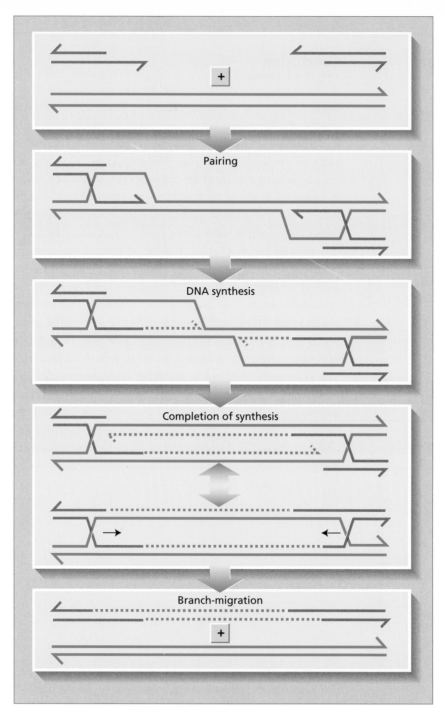

Fig. 2.35 Double-strand break repair (DSBR) by branch-migration. After Thaler *et al.*, 1987. In this version of the model, pairing occurs to form two Holliday junctions and two replication forks. The Holliday junctions are not resolved and replication results in the formation of a double Holliday junction structure without strand-transfer taking place. Catalysed branch-migration of the two junctions towards each other will resolve the intermediate into recombinant products. Since strand-transfer has not occurred, only non-crossover products are possible. Branch-migration and DNA synthesis may occur in concert with each other and may be facilitated by helicase and topoisomerase activities.

Further reading

Kowalczykowski S. C., Dixon D. A., Eggleston A. K., Lauder S. D. & Rehrauer W. H. (1994). Biochemistry of homologous recombination in *Escherichia coli*. *Microbiol. Revs.* **58**, 401–465.

Kucherlapati R. & Smith G. R. (1988). *Genetic Recombination. A Collection of Review Articles on Homologous Recombination*. American Society for Microbiology, Washington.

Low K. B. (1988). *The Recombination of Genetic Material. A Collection of Review Articles on Homologous Recombination*. Academic Press, San Diego.

Myers R. S. & Stahl F. W. (1994). *Chi* and the RecBCD enzyme of *Escherichia coli. Ann. Rev. Genet.* **28**, 49–70.

Petes T. D., Malone R. E. & Symington L. S. (1991). Recombination in Yeast. In: *The Molecular and Cellular Biology of Yeast Saccharomyces. Cold Spring Harbour Press* **1**, 407–521.

Stahl F. W. (1979). *Genetic Recombination, Thinking About it in Phage and Fungi*. Freeman, USA.

Whitehouse H. L. K. (1982). *Genetic Recombination, Understanding the Mechanisms*. Wiley, New York.

References

Amati P. & Meselson M. S. (1965). Localised negative interference in bacteriophage lambda. *Genetics* **51**, 369–379.

Cox M. M. & Lehman I. R. (1981). Directionality and polarity in RecA protein-promoted branch-migration. *Proc. Natl. Acad. Sci. USA* 78, 6018–6022.

Hagemann A. T. & Rosenberg S. M. (1991). Chain bias in chi-stimulated heteroduplex patches in the λ *ren* gene is determined by the orientation of λ *cos*. *Genetics* **129**, 611–621.

Kahn R., Cuningham R. P., Das Gupta C. & Radding C. M. (1981). Polarity of heteroduplex formation promoted by *Escherichia coli* RecA protein. *Proc. Natl. Acad. Sci. USA* **78**, 4786–4790.

Kobayashi I., Murialdo H., Crasemann J. M., Stahl M. M. & Stahl F. W. (1982). Orientation of cohesive end site (cos) determines the active orientation of *Chi* in stimulating RecA-RecBCD mediated recombination in lambda lytic infections. *Proc. Natl. Acad. Sci. USA* **79**, 5981–5985.

Kobayashi I., Stahl M. M. & Stahl F. W. (1985). The mechanism of the *Chi*–cos interaction in RecA–RecBC-mediated recombination in phage lambda. *Cold Spring Harbour Symp. Quant. Biol.* **49**, 497–506.

Kobayashi I., Stahl M. M., Leach D. R. F. & Stahl F. W. (1983). The interaction of cos with *Chi* is separable from DNA packaging in recA-recBC-mediated recombination of bacterio-phage lambda. *Genetics*, **104**, 549–570.

Kowalczykowski S. C., Dixon D. A., Eggleston A. K., Lauder S. D. & Rehrauer W. H. (1994). Biochemistry of homologous recombination in *Escherichia coli*. *Microbiol. Revs.* **58**, 401–465.

Meselson M. S. & Radding C. M. (1975). A general model for genetic recombination. *Proc. Natl. Acad. Sci. USA* **72**, 358–361.

Meselson M. S. & Weigle J. J. (1961). Chromosome breakage accompanying genetic recom-bination in bacteriophage. *Proc. Natl. Acad. Sci. USA* **47**, 857–868.

Meselson M. S. (1965). The Molecular Basis of Genetic Recombination. *Mendel Centennial Symposium*.

Murray N. E. (1960). Complementation and recombination between methionine-2 alleles in *Neurospora crassa. Heredity* **15**, 207–217.

Murray N. E. (1963). Polarized recombination and fine structure within the me-2 gene of

Neurospora crassa. *Genetics* **48**, 1163–1183.

Paquette N. & Rosignol J.-L. (1978). Gene conversion spectrum of 15 mutants giving post-meiotic segregation in the b2 locus of Ascobolus immersus. *Molec. Gen. Genet.* **163**, 313–326.

Potter H. & Dressler D. (1977). On the mechanism of genetic recombination, the nature of recombination intermediates. *Proc. Natl. Acad. Sci. USA* **74**, 4168–4172.

Rao B. J. & Radding C. (1993). Homologous recognition promoted by RecA protein via non-Watson–Crick bonds between identical DNA strands. *Proc. Natl. Acad. Sci. USA* **90**, 6646–6650.

Siddiqui I., Stahl M. M. & Stahl F. W. (1991). Heteroduplex chain polarity in recombination of phage λ by the Red, RecBCD, RecBC(D−) and RecF pathways. *Genetics* **128**, 7–22.

Signer E. R. & Weil J. (1968). Recombination in bacteriophage lambda I. Mutants deficient in general recombination. *J. Molec. Biol.*, **34**, 261–271.

Stahl F. W. & Stahl M. M. (1975). rec-mediated recombinational hot spot activity in bacteriophage lambda IV. Effect of heterology on *Chi*-stimulated crossing over. *Molec. Gen. Genet.* **140**, 29–37.

Stahl F. W., Kobayashi I. & Stahl M. M. (1985). In phage lambda, cos is a recombinator in the Red pathway. *J. Molec. Biol.* **181**, 199–209.

Stahl F. W., McMilin K. D., Stahl M. M., Crasemann J. M. & Lam S. (1974). The distribution of crossovers along unreplicated lambda bacteriophage chromosomes. *Genetics* **77**, 395–408.

Stahl F. W., Stahl M. M., Malone R. E. & Crasemann J. M. (1980). Directionality and non-reciprocallity of *Chi*-stimulated recombination in phage lambda. *Genetics* **94**, 235–248.

Szostak J. W., Orr-Weaver T. L., Rothstein R. J. & Stahl F. W. (1983). The double-strand break repair model for conversion and crossing over. *Cell* **33**, 25–35.

Taylor A. F. (1988). RecBCD enzyme of *Escherichia coli*. In Kulcherapati R. & Smith G. R. (eds) *Genetic Recombination*, pp. 231–263. American Society for Microbiology, Washington.

Thaler D.S., Stahl M. M. & Stahl F. W. (1987). Tests of the double-strand-break repair model for Red-mediated recombination of phage λ and plasmid λdv. *Genetics* **116**, 501–511.

Weil J. & Signer E. R. (1968). Recombination in bacteriophage lambda II. Site-specific recombination promoted by the integration system. *J. Molec. Biol.* **34**, 273–279.

West S. C., Casuto E. & Howard-Flanders P. (1981). Heteroduplex formation by RecA protein, polarity of strand exchanges. *Proc. Natl. Acad. Sci. USA* **78**, 6149–6153.

3 The Roles of Homologous Recombination

3.1 Introduction

Genetic recombination affects the balance between fitness and flexibility. A flexible genome allows a rapid response to selection in a variable environment. However, in a stable environment, an organism will be at a disadvantage if its favourable combinations of alleles are broken up to produce variable offspring.

There are other consequences of homologous recombination. The most important of these is the repair of DNA that has been damaged on both strands, but homologous recombination can also mediate programmed DNA rearrangements such as mating-type switching or antigenic variation, and the spread of selfish DNA sequences. Furthermore, it assists accurate chromosome segregation and can contribute to DNA replication.

3.2 Genetic mixing

Genetic mixing (or mixis) is the recombination of genetic information present in different individuals that is facilitated by sexual reproduction. There are two contributions of recombination to mixis. The first is the independent assortment of maternal and paternal chromosomes in meiosis. The organization of the genetic material into several chromosomes ensures that not all genes are linked. In a diploid organism, maternal and paternal chromosomes are segregated at random to the gametes ensuring a wide variety of genotypes. The second is homologous recombination which allows alleles of genes that are linked to be separated or brought together during meiosis. These concepts are introduced in Chapter 1.

Recombination has the four possible effects described below. Two of these will lead to selection for recombination and two against.

1 Combination of favourable alleles

In the early 1930s, R. A. Fisher and H. J. Muller suggested that recombination might increase the rate of evolution by bringing together combinations of favourable alleles more rapidly than can occur by mutation alone. Imagine two members of a species each of which acquires a different mutation that results in increased fitness. Without recombination, the combination of the two favourable mutations in a single individual amongst the progeny of these individuals, will depend on further mutation. On the other hand, with recombination, this outcome can arise more rapidly by genetic exchange (Fig. 3.1).

2 Disruption of favourable alleles

The converse of the Fisher–Muller argument is that when combinations of

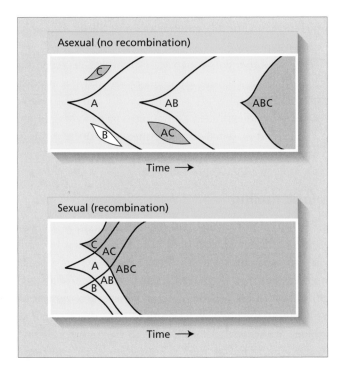

Fig. 3.1 The effect of genetic recombination on the rate of combination of favourable alleles in evolution. After Maynard-Smith (1989) (By permission of Oxford University Press.). Without recombination, the establishment of a favourable combination of alleles will depend on mutation alone. Each favourable mutation will have to occur in a clone carrying another favourable mutation. On the other hand, with recombination, mutations that have arisen in independent clones can be combined within the same individual. This accelerates the formation of novel combinations of genotypes and therefore the response to selection.

favourable alleles have evolved, recombination will disrupt them. This will lead to selection against recombination.

3 Avoidance of deleterious alleles

All organisms must carry a burden of deleterious alleles caused by mutation and there are two ways of avoiding the accumulation of these alleles. The first is back-mutation which is very slow and the second is recombination between individuals carrying different deleterious mutations to regenerate progeny lacking them. This second mechanism is significantly more rapid. In the absence of recombination (or back-mutation) a species has no escape from a ratchet that leads to the continuous accumulation of deleterious alleles due to the chance elimination of the class of individuals carrying the smallest number of deleterious alleles and the absence of a mechanism for its regeneration. This was originally suggested by H. J. Muller in 1964 and has come to be known as Muller's ratchet (Fig. 3.2).

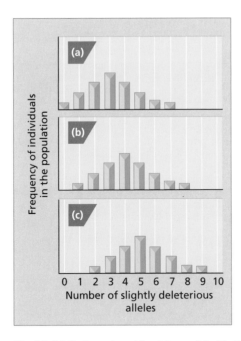

Fig. 3.2 Muller's ratchet. After Maynard-Smith (1989). We can imagine a population of individuals with a distribution of slightly deleterious alleles caused by mutation. For the sake of argument, we will assume that initially (as shown in a) the population includes a small number of individuals with no such alleles. At a future date there is a chance that all these individuals, despite their high fitness, leave no offspring. The distribution will change so that the class with the least number of deleterious alleles becomes that with one (as shown in b). The ratchet has turned once and will then keep turning as the class with the least number of deleterious mutations is lost by chance (see c).

4 Combination of deleterious alleles

Just as recombination can reconstitute a genome without deleterious alleles, it can create combinations of such alleles, and this will lead to selection against recombination.

How can we untangle the evolutionary advantages and disadvantages of genetic mixing? A key is to understand the effect of recombination on linkage disequilibrium. Linkage disequilibrium is a measure of the association of alleles at two loci as defined in Fig. 3.3. If alleles at two loci are in linkage equilibrium, recombination just maintains the status quo because it undoes as many new combinations as it creates. On the other hand, if alleles are in linkage disequilibrium, recombination acts to accelerate the movement of gene frequencies towards linkage equilibrium. It shuffles the pack of cards. If favourable combinations of alleles are less common than at linkage equilibrium, there will be selection for recombination. If favourable combinations are more common there will be selection against recombination.

Linkage disequilibrium

If the probability of finding an allele (A) at a particular locus does not depend on the presence or absence of an allele (B) at another locus, the chance of finding a gamete with both A and B alleles is simply the product of the individual probabilities:

$$p_{AB} = p_A p_B$$

However, if these probabilities are not independent, this product will not hold and:

$$p_{AB} = p_A p_B + D$$

where D is the coefficient of linkage disequilibrium and measures the departure from random association of alleles A and B. Similarly, if there are alternative alleles a and b at these loci, the probabilities of the four different gametes will be:

$$
\begin{aligned}
p_{AB} &= p_A p_B + D \\
p_{Ab} &= p_A p_b - D \\
p_{aB} &= p_a p_B - D \\
p_{ab} &= p_a p_b + D
\end{aligned}
$$

or:

$$D = p_{AB} p_{ab} - p_{Ab} p_{aB}$$

D is the difference between the product of the gametic allele probabilities in coupling (*AB* and *ab*) and in repulsion (*Ab* and *aB*). By convention, alleles contributing to a particular phenotype are denoted by the same type of symbol (e.g. A and B both contribute to increased height and a and b contribute to decreased height).

Fig. 3.3 Linkage disequilibrium.

Linkage disequilibrium can occur as a consequence of random effects due to finite population size and to consequences of selection. In fact linkage disequilibrium conferring selective advantage can change to linkage disequilibrium conferring disadvantage if the fitnesses of particular alleles change as a consequence of new environmental conditions. The discussion of these effects lies beyond the scope of this book but readers are referred to Michod & Levin (1988).

Unlinked genes can recombine freely with each other but linked genes recombine only as a function of their chromosomal separation and the homologous recombination mechanisms available. Homologous recombination rates vary considerably between organisms and between different chromosomal regions in one organism. These rates are under the control of genes and sites that often act locally and can be altered by selection for high or low recombination. The local action of these recombinators allows flexibility in the alteration of recombination rates to suit particular genes that happen to be under selection without affecting other regions and also allow the recombination genes to hitch-hike with the regions they affect.

3.3 ## Chromosome segregation

In meiosis, homologous chromosomes must pair in order to ensure that one copy of each segregates to the gametes. Meiotic recombination occurs after the chromosomes have first replicated and leads to the formation of chiasmata which are points of crossing over between chromatids and are visible under the light microscope (see Figs 1.5 and 1.6). These chiasmata determine the proper alignment of bivalents on the first metaphase plate (as shown in Fig. 1.5) and are essential for the proper segregation of most chromosomes. For most chromosomes there is a rule that at least one chiasma forms per bivalent and in meioses, where chiasmata are prevented, chromosomes fail to segregate properly.

Nevertheless, there are some special situations where correct segregation occurs in the absence of recombination. For instance, no homologous recombination occurs in the meioses of male *Drosophila* or female *Lepidoptera,* and the small 4th chromosome of *Drosophila* does not undergo homologous meiotic recombination in females either. It is believed that an additional pairing system exists that allows homologous chromosomes to align without necessarily undergoing recombination. This system may involve heterochromatic pairing. Nevertheless, the fact that recombinational exchange does affect segregation is underlined by the behaviour of *nod* mutants of *Drosophila melanogaster.* They specifically disrupt the segregation of chromosomes not undergoing homologous recombination, while chromosomes with chiasmata segregate properly.

3.4   **DNA repair**

Homologous recombination is essential for the accurate repair of DNA double-strand gaps. Double-strand gaps can form by exonucleolytic attack at the site of a double-strand break. Because information is lost from both strands of the DNA, the correct sequence can only be retrieved from another duplex. The double-strand break repair (DSBR) model (see Chapter 2) can be applied to the repair of damage caused by sources such as ionizing radiation and hydroxyl radicals. This is illustrated in Fig. 3.4. Diploid organisms have the opportunity to repair their double-strand gaps with information present on a homologous chromosome or that present on a sister chromatid if damage has occurred after DNA replication. There is also a choice as to whether the repair is associated with crossing over or not. Haploid organisms have only the possibility to repair damage by recombination with a sister double-strand. These interactions and their consequences are outlined in Figs 3.5 and 3.6.

Homologous recombination can also facilitate replication past the site of DNA damage. This is not repair, since the lesion (e.g. a pyrimidine dimer) is not removed but promotes cell survival and repair of the damage by other pathways. A model for this role of recombination was originally suggested by Paul Howard-Flanders and described as 'post-replication repair'. The actual mechanism of this type of damage avoidance is not clear but a model is depicted in Fig. 3.7.

3.5   **Mating-type interconversion**

The interconversion between different mating-types has been identified in several fungal species but has been studied most extensively in the yeast *Saccharomyces cerevisiae*. Haploid yeast cells switch in a controlled way between the two mating types 'a' and 'α' which can then mate to produce a/α diploids. This switching involves four primary loci: *MAT*, the mating-type locus that can either contain a or α information and which defines the cell's mating-type; *HMR*, that normally contains a silent (non-expressed) copy of a; *HML* that normally contains a silent copy of α; and *HO*, that controls switching. Switching occurs by recombination of the silent copies, *HMR* and *HML*, with *MAT* to alternate the information contained at the mating-type locus. The *HM* loci and *MAT* share homologous regions W, X and Z that flank the allele-specific Y region: Ya or Yα (Fig. 3.8).

One of the first indications that switching occurs by recombination was the remarkable finding made in the laboratories of Don Hawthorne and Ira Herskowitz that mutations at MAT can be 'cured' to wild-type. This demonstrated that there exists a repository of wild-type information that

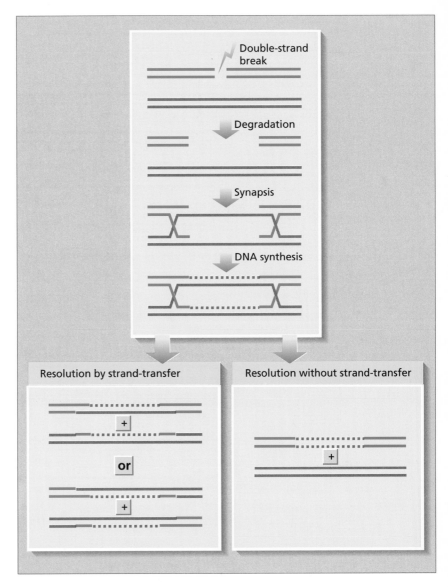

Fig. 3.4 The repair of DNA double-strand breaks and gaps. When a double-strand break is subjected to exonucleolytic degradation, a gap is formed where coding information has been lost. The degradation can either proceed more rapidly in a 5'–3' or a 3'–5' direction, depending on the system. This results in the formation of single-stranded overhangs. However, since the figure illustrates the general principles that apply in either case, no such asymmetry is indicated. As described in Chapter 2, a 5' overhang is the primary product of RecBCD nuclease of *E. coli* but a 3' overhang is the primary product of the 5'–3' exonucleases of *S. cerevisiae*. Synapsis then occurs and DNA synthesis replaces the missing information. The products are then completed, either by strand-transfer leading to splice and patch recombinants, or without strand-transfer ensuring that no crossing over takes place.

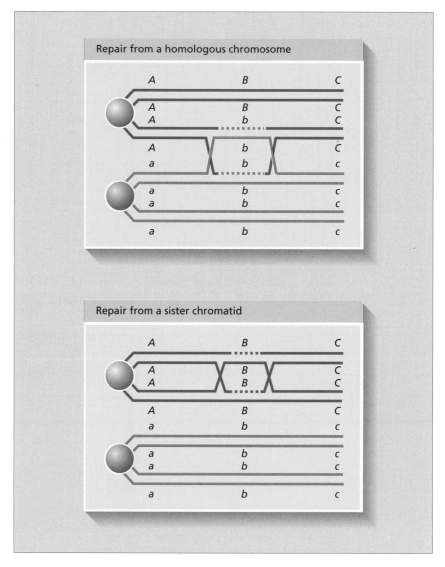

Fig. 3.5 Consequences of DNA double-strand break repair (DSBR) in a diploid organism.
Repair from a homologous chromosome is not the same as repair from a sister chromatid.
Since homologous chromosomes carry different alleles at many loci, repair can cause
recombination. This can either be in the form of crossing over, as shown between genes *A/a*
and *C/c* in the figure, or as non-reciprocal recombination, that may or may not be
associated with crossing over, as shown between *B/b* and *A/a* or *C/c*. On the other hand
repair from a sister chromatid does not cause recombination since sisters are identical. Only
the central intermediate in DSBR is illustrated (see Fig. 3.4 for the possible ways of
resolving this structure).

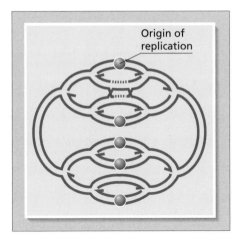

Fig. 3.6 Double-strand break repair (DSBR) in a haploid organism with a circular chromosome. In a haploid organism, DSBR can only occur from a sister chromosome or a partially replicated sister chromosome. In rapidly growing bacterial cells, multifork replication ensures that several copies of most, if not all, genes are present in the same cell. Repair in such a replicating chromosome is shown.

can 'repair' the mutant information. The cloning of the *MAT* and *HM* loci has confirmed that the *HM* loci carry silent copies of mating-type information. The switching itself is under the control of the *HO* gene which encodes an endonuclease that specifically cleaves the junction between the allele-specific Y region and the homologous Z1 region of the *MAT* locus. This has led to the proposal that mating-type interconversion occurs by DSBR. A specific version of the DSBR model that accounts for several other properties of the reaction is shown in Fig. 3.9.

3.6   Antigenic variation

A number of pathogenic organisms attempt to evade the immunological defences of their hosts by varying their surface proteins during the course of an infection. Two of the best studied systems of antigenic variation are those found in *Trypanosoma* and *Neisseria*.

The African Trypanosomes cause mammalian parasitic infections including the sleeping sickness that is spread by the tsetse fly. *Trypanosoma brucei brucei,* which is closely related to the sleeping sickness pathogen but is non-infectious in humans, has been most extensively studied. *T. equiperdum,* which infects horses, has also proved an interesting model system. Following infection, waves of parasitemia occur as parasites with one variant surface glycoprotein (VSG) first multiply and then are destroyed by the host immune system, only to be replaced with derivatives carrying a second VSG. This replacement of one VSG by another can continue for many cycles and one experimental rabbit is reported to have been the host to over 100 different antigenic types.

As a rule, only one VSG gene is expressed at one time and this gene is

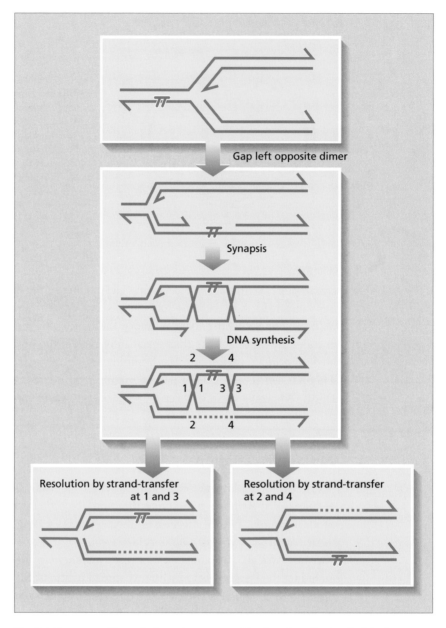

Fig. 3.7 Damage avoidance by homologous recombination. A replication fork is shown, heading for a pyrimidine dimer (π). The dimer is not recognized as a good template and a gap is left opposite it. As it is, this gap cannot be faithfully filled in because of the dimer. However the single-stranded DNA acts as a signal to initiate homologous recombination. The strand with the dimer becomes paired to an intact sister-strand and the good strand identical to that with the dimer is copied (DNA synthesis). Resolution by strand-transfer will then result in the two structures illustrated. Strand-transfer at 1 and 3 leaves the dimer on a newly-replicated strand; at 2 and 4 on the parental strand; at 1 and 4, and 2 and 3 at junctions between newly replicated and parental strands. Note that no repair has taken place; the dimer remains present. However other systems now have time to accurately repair the damage.

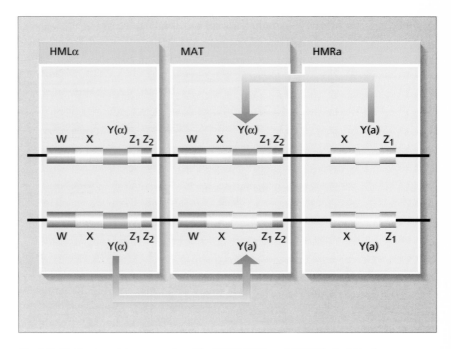

Fig. 3.8 Mating-type interconversion. The *MAT*, *HML* and *HMR* loci of *Saccharomyces cereviciae* contain three blocks of homologous DNA sequences, W, X and Z, within which recombination can occur, flanking region Y which specifies the mating type 'a' or 'α'. The mating-type of the cell is determined by the Y DNA at the *MAT* locus which alternates between the two types (Ya and Yα).

called the expression-linked copy or ELC. It is always located next to a telomere but linkage to a telomere is not sufficient to allow activation since there also exist many telomere linked copies that are silent. The number of these genes is not known but may be as high as 200 since there are approximately 100 chromosomes. Many of these are mini-chromosomes whose only role may be to carry VSG genes. In addition to telomere-linked genes there are interior-located genes and parts of genes, all of which are silent too. This vast repository of silent information (approximately one tenth of the genome) is called upon by recombination to modify the ELC and thereby allow the parasite to escape the host's immune system. The organization of the ELC is shown in Fig. 3.10. Antigenic-switching occurs by recombination of the ELC with silent copies via a mechanism similar to that proposed for mating-type interconversion (shown in Fig. 3.9). In fact, recombination with several silent copies and part copies can further increase the potential repertoire of VSGs. These more complex recombination events are not detected until late in infection since they occur at a lower frequency than the simple events.

Neisseria gonorrhoeae is the pathogen responsible for the human

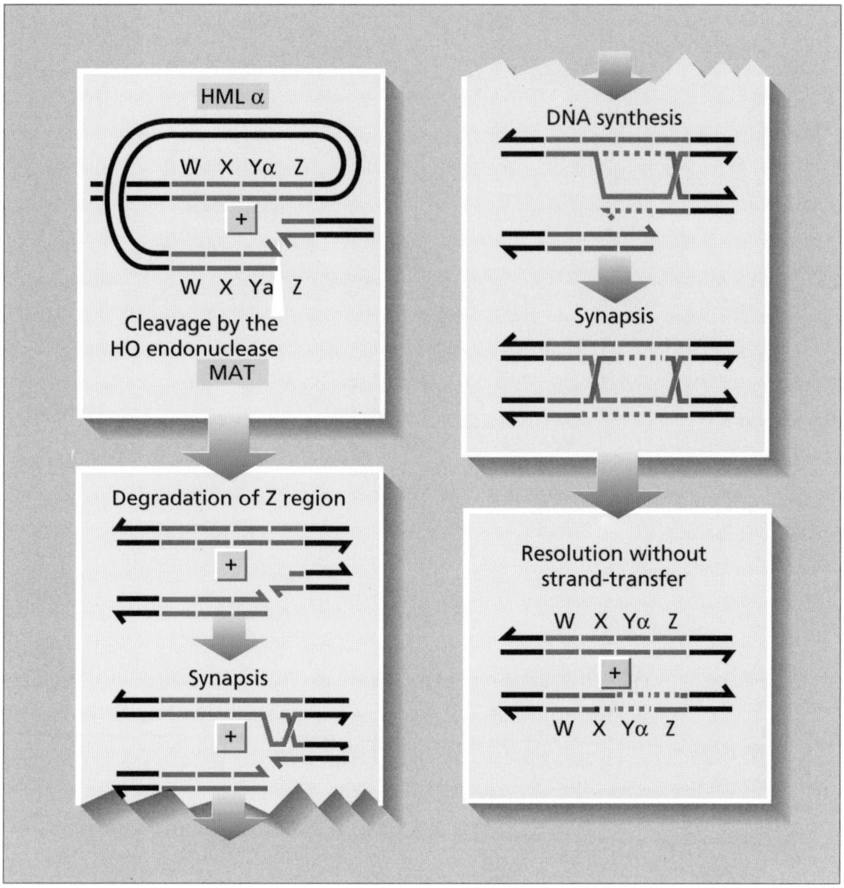

Fig. 3.9 A model for mating-type interconversion. The HO endonuclease is known to recognize and cleave the DNA sequence at the Y/Z junction. Degradation of Z DNA then occurs on the 5′ strand to leave a 3′ single-stranded tail. The Y region remains insensitive to degradation for a significant time period. Recombination is initiated in the Z region and a hypothetical intermediate containing one Holliday junction is formed. DNA synthesis and the formation of a second Holliday junction may then occur in the X region. Genetic evidence indicates that heteroduplex DNA exists in X and that the formation of the junction involved invasion of a 3′ end (as in Z). The resolution of this structure must then occur according to specific rules. However, the details of resolution are not yet clear. Resolution may occur by strand-transfer or, as shown here, without strand-transfer. The attraction of resolution without strand-transfer, as originally proposed by Frank Stahl in Naysmith (1982), is that no heteroduplex DNA persists at the HM locus and no crossing over occurs. This is consistent with the observed unidirectional transfer of information from HM to MAT and the absence of crossing over.

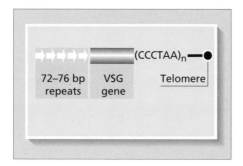

Fig. 3.10 Organization of the expression-linked copy of the VSG gene of *Trypanosoma brucei*. The expression-linked copy of the VSG gene is always located close to a telomere and is surrounded on either side by barren regions of 5–20 kb of repeated DNA. Upstream of the gene is a region composed of large numbers of 72–76 bp repeats that are also found in smaller numbers adjacent to internal non-expressed copies of the gene. They may provide the recognition signal for the recombination event involved in switching. Downstream of the gene, is another 5–20 kb region consisting of $(CCTAA)_n$ repeats interspersed with AT-rich sequences which are followed by the telomere.

sexually transmitted disease, gonorrhoea. The bacterium responsible is relatively fragile but escapes the immune system by variation of two of its cell-surface proteins, pilin and P.II.

Pilin is the protein constituent of the pili which are required for cell-surface adhesion and therefore the pathogenicity of the bacterium. It has a conserved 51 amino-acid amino-terminus and a variable 110 amino-acid carboxy-terminus. Within this carboxy-terminus there are six short conserved regions. The gonococcal genome contains one complete, expressed copy of the pilin gene and 12–20 partial, silent copies clustered at several loci surrounding the complete gene. The silent copies encode the carboxy-terminal part of the protein and are the repository of variation that is available to the expressed copy by recombination (Fig. 3.11). As for the cases of mating-type interconversion and VSG switching, the outcome of recombination is the non-reciprocal transfer of information from the silent copies to the expressed copy.

The P.II protein is a major component of the outer membrane and is also subject to variation. However, the mechanism involves illegitimate rather than homologous recombination.

3.7   Intron homing and intron loss

Many eukaryotic, and some prokaryotic, genes are divided into coding regions (exons) and intervening sequences (introns) that are non-coding and must be removed form the messenger RNA before translation. The removal

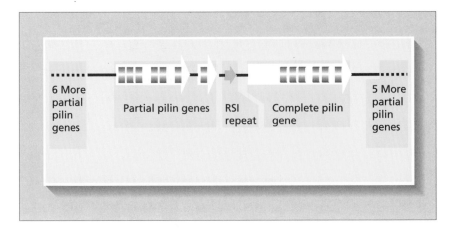

Fig. 3.11 Organization of the pilin locus of *Neisseria gonorrhoeae*. The organization
of the pilin genes in a strain of *Neisseria gonorrhoeae* is illustrated. The whole locus is
interspersed with a common 39–40 bp repeat, denoted RS1, that is often, but not always,
located at the downstream end of a partial pilin gene. The shaded regions represent base
sequences common to all complete and partial genes.

of introns from RNA occurs by a reaction known as splicing. This is an
'RNA-recombination' reaction which lies outside the scope of this book
however there are also DNA-recombination reactions that affect introns.

One class of introns has been designated 'Group I' and is characterized
by an ability to self-splice in the presence of guanine. These introns are, in
fact, ribozymes (autocatalytic RNAs) which sometimes encode a maturase
to assist splicing. A sub-class of these introns are mobile. They can move
to an intron-free copy of their gene by homologous recombination. This is
called homing. To facilitate this reaction, each of these introns encodes an
endonuclease that recognizes a DNA sequence in the intron-free gene.
Cleavage at this site stimulates double-strand break repair (DSBR) and
copying of the intron as shown in Fig. 3.12. The mechanism of this reaction is likely to be similar to the other DSBR reactions described in this
chapter. The product of homing has its endonuclease recognition-site split
by the newly inserted intron and is therefore no longer a substrate for
cleavage.

The endonuclease gene itself is the primary element that is mobile and
the intron is a safe home where it does not disrupt gene function. Not all
mobile endonuclease genes are located in introns. An unusual mobile
endonuclease gene is located in the coding sequence of the VMAI gene of
Saccharomyces cereviciae. Here the endonuclease is safe because its protein sequence is processed out of the VMAI protein precursor after its
translation.

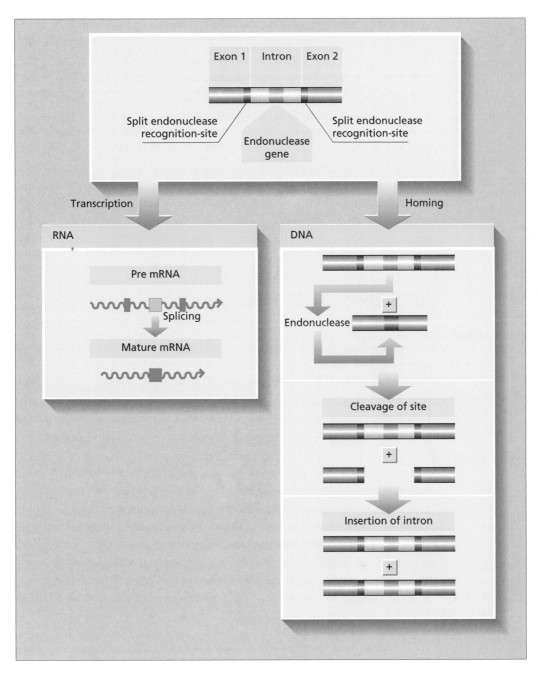

Fig. 3.12 Intron homing. If an intron-free gene is present in the same cell as its homologue carrying a mobile intron, the endonuclease gene will cleave the intron-free copy and DSBR will result in copying the intron to the cleaved site.

Mobile introns have, so far, only been detected in bacteriophage T4 of *Escherichia coli*, the mitochondrial DNA of *Saccharomyces cerevisiae* and *Chlamydomonas eugamentos* and the nuclear DNA of *Physarum polycephalum*. The significance of this restriction to prokaryotes and lower eukaryotes (and the preponderance of mobile introns in the DNA of mitochondrial organelles) remains to be determined.

Although these specific introns have clearly evolved to be mobile and their endonuclease genes can be considered as selfish DNA sequences, all introns have the potential to spread or be removed by homologous recombination between chromosomes that do or do not contain them. Furthermore, recent evidence has been obtained that intron-loss can occur by recombination between the reverse-transcribed copy of a mature (spliced) mRNA and the chromosome. These interactions are summarized in Fig. 3.13.

3.8   DNA replication and packaging

The primary function of the homologous recombination in bacteriophages is DNA replication. The two most extensively studied systems are those of bacteriophages T4 and λ.

Bacteriophage T4 encodes several recombination genes including its own '*recA*' gene *uvsX*. Giesella Mosig and co-workers have extensively studied the role of recombination in T4 DNA replication and a model of how recombination leads to the formation of replication forks is shown in Fig. 3.14.

The recombination proteins of bacteriophage λ are encoded by the *redα* and *redβ* genes described in Chapter 2. Their function is to stimulate the formation of rolling-circles by recombination between linear and circular molecules (Fig. 3.15). This has two outcomes. Firstly rolling-circle replication is a very efficient mechanism of producing daughter molecules and secondly these progeny are linked in a long concatamer which is the ideal precursor for packaging of the DNA into the phage head.

This appears to be a role of recombination adopted by specialized

Fig. 3.13 Loss and gain of introns by recombination. Theoretically, the acquisition of an intron by an intron-free copy of a gene does not necessitate a mobile intron. Homologous recombination between two chromosomes catalysed by random DNA-breakage or specific recombinators should result in such an outcome at a low frequency. Similarly, homologous recombination should occasionally result in intron-loss. No examples of this type of intron-loss or gain have yet been reported, but intron-loss as a consequence of transcription, splicing, reverse transcription and recombination has been detected in model systems. This documented route for intron-loss is shown with red arrows. Exons are shaded and the

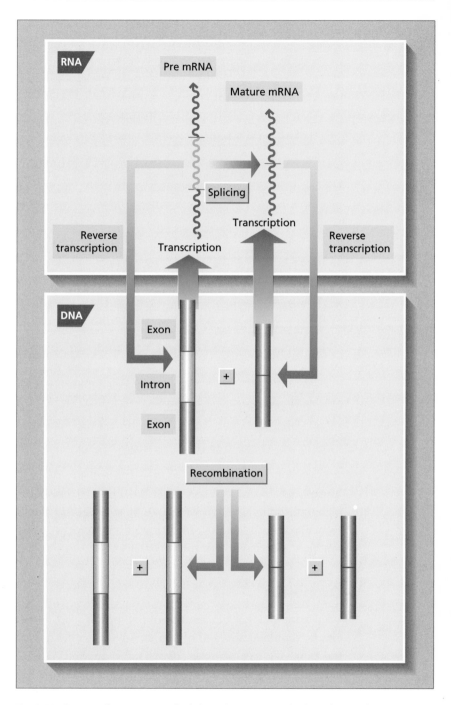

Fig. 3.13 (*Continued*) intron is unshaded. In the centre are the homologous chromosomes of a diploid that is heterozygous for an intron. Transcription can lead to both pre-mRNA and mature mRNA which can be reverse transcribed back into DNA. Recombination between the chromosomal sequences themselves, or between the chromosomal sequences and the reverse-transcribed copies, can potentially result in intron-loss or gain.

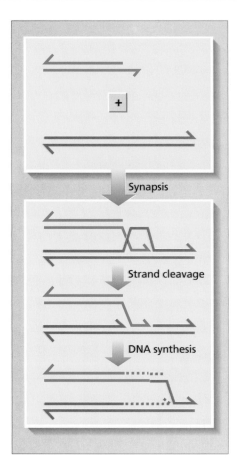

Fig. 3.14 Formation of replication forks by T4 recombination. Recombination that is initiated by an invading 3′ end is proposed to generate new replication forks as shown. This results in the formation of multiple-branched chromosomes during T4 replication.

'organisms' (bacteriophages) that are under strong selective pressure for high replication rates. However, we must remember that most homologous recombination reactions investigated do have a component of DNA synthesis associated with the exchange.

3.9   Conclusion

We cannot look back thousands of millions of years to the origin of homologous recombination and obtain an answer to the question of its primary function. We can however make some guesses. Originally, when DNA became established as a coding material in a previously RNA world, recombination may have evolved as a mechanism to permit the genome to expand by picking up other pieces of DNA. Recombination may also have contributed to the replication strategies of these primitive DNAs. What we see today as minor roles of recombination in the propagation of some introns and bacteriophages may be pointers to the most ancient functions

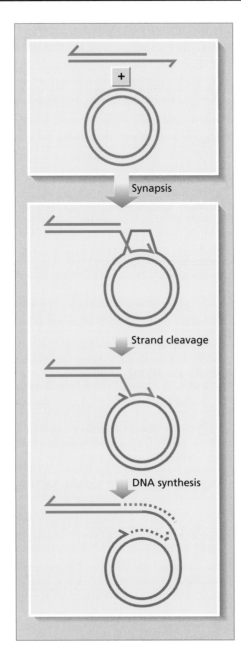

Fig. 3.15 Formation of rolling-circles by λ recombination. If the same sequence of events as described in Fig. 3.14 occurs between a linear and a circular molecule, the result is a rolling-circle instead of a branch.

of recombination. When genomes had become larger and more stable, the importance of repair of double-strand breaks and gaps may have taken over as the major function of recombination. DNA repair remains an important role, however homologous recombination is also used for the regulation of gene expression as demonstrated by mating-type interconversion and antigenic variation. Finally, homologous recombination in meiosis and sexual reproduction facilitate the breakdown of linkage disequilibrium in organisms with large genomes. At present, it is likely that selection operates on all these functions of homologous recombination. As organisms have become more complex, their recombination functions have become more specialized. It is, therefore, not surprising that, for instance in *Saccaromyces cerevisiae*, there are some genes affecting mitotic, meiotic or mating-type interconversion individually and others that affect all three.

Further reading

Belfort M. (1990). Phage T4 introns: self-splicing and mobility. *Ann. Rev. Genet.* **24**, 363–385.

Donelson J. E. (1989). DNA rearrangements and antigenic variation in African Trypanosomes. In Berg D. E. & Howe M. M. (eds) *Mobile DNA*, pp. 763–782. American Society for Microbiology, Washington.

Friedberg E. (1985). *DNA repair*. Freeman, New York.

Haber J. E. (1992). Mating-type gene switching in *Saccharomyces cerevisiae*. *Trends Genet.* **8**, 446–452.

Hawley R. S. (1988). Exchange and chromosomal segregation in eukaryotes. In Kulcherlapati R. & Smith G. R (eds) *Genetic Recombination*, pp. 497–528. American Society for Microbiology, Washington.

Klar A. J. S. (1989). The interconversion of yeast mating type: *Saccharomyces serevisiae* and *Schizosaccharomyces pombe*. In Berg D. E. & Howe M. M. (eds) *Mobile DNA*, pp. 671–692. American Society for Microbiology, Washington.

Maynard-Smith J. (1989). *Evolutionary Genetics*. Oxford University Press, Oxford.

Michod R. E. & Levin B. R. (1988). *The Evolution of Sex; an Examination of Current Ideas*. Sinauer, Sunderland.

Strathern J. N. (1988). Control and execution of homothalic switching in *Saccharomyces cerevisiae*. In Kulcherlapati R. & Smith G. R (eds) *Genetic Recombination*, pp. 445–464. American Society for Microbiology, Washington.

Swanson J. & Koomey M. (1989). Mechanisms for variation of pili and outer membrane protein II in *Neisseria gonorrhoeae*. In Berg D. E. & Howe M. M. (eds) *Mobile DNA*, pp. 743–762. American Society for Microbiology, Washington.

References

Maynard Smith J. (1989). *Evolutionary Genetics*. Oxford University Press, Oxford.

Naysmith K. A. (1982). Molecular genetics of yeast mating type. *Ann. Rev. Genet.* **16**, 439–500.

4 Site-Specific Recombination

4.1 Introduction

Site-specific recombination, as its name suggests, involves the interaction of specific DNA sites. It is also distinguished from homologous recombination by the mechanism of recognition used by the two recombining partners. In homologous recombination, one DNA sequence recognizes another (admittedly with the help of proteins such as RecA) while here the proteins themselves mediate recognition (Fig. 4.1). The recombining sites may share homology by virtue of their interaction with the recombinase and/or steps in the recombination reaction, but the homology is not involved in recognition. The two best studied model systems are bacteriophage λ integration/excision and the resolution of co-integrates by Tn3 resolvase. This chapter, therefore, concentrates on mechanistic studies of these two reactions and then illustrates the biology of site-specific recombination with examples of related systems.

Elucidation of the mechanisms of site-specific recombination has been greatly facilitated by the topological analysis of the substrates and products of the reaction. This chapter, therefore, starts with a brief review of the essential topological and conformational features of DNA needed to understand this work.

4.2 DNA conformation and topology

We all know that DNA is a double-helix but may not all understand what is meant by the description of this double-helix as right-handed. Helices can be either right or left-handed as shown in Fig. 4.2 and the normal helicity of DNA (B-DNA) happens to be right-handed. The B-DNA helix has approximately 10.4 base pairs (bp) per turn when relaxed, but can be wound-up or unwound to more or less bp per turn. These stressed conformations must be anchored in some way to prevent their return to

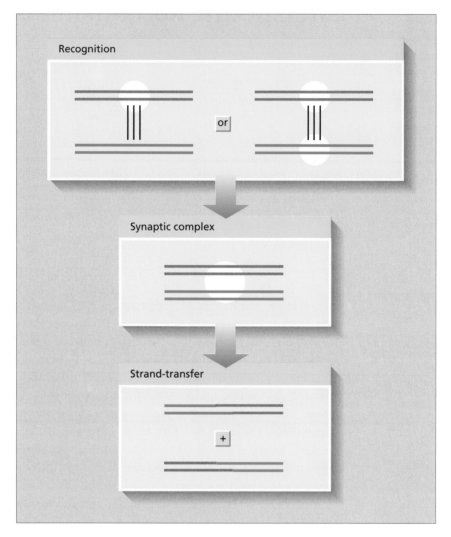

Fig. 4.1 Site-specific recombination. The recombinase or recombinase complex recognizes a specific DNA sequence and binds. It then either interacts with another DNA sequence which may or may not be identical to the first and may or may not have pre-bound recombinase. A synaptic complex is formed within which the strand-transfer reactions take place. In all the examples of site-specific recombination where the information is available, the reaction proceeds via a covalent protein–DNA intermediate and results in small regions of heteroduplex of between 2 and 8 bp in length. As for homologous recombination, four strand-transfer reactions must take place to complete the recombinant.

the relaxed 10.4 bp per turn. This is the case in a closed circular DNA molecule. The anchoring provided by closing the circle only comes in discrete steps since the strands must wrap around each other an integral number of times. Figure 4.3 represents a double-helix with three, four or five turns. These are called topoisomers because they are topologically distinct.

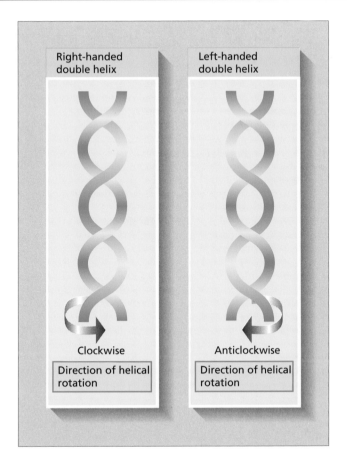

Fig. 4.2 Right- and left-handed helices. There are two types of helix that can be formed. These are called right-handed and left-handed, as defined in this figure. A right-handed helix has the same (clockwise) rotation as a corkscrew, or the rotation required to insert a screw. Note that the right- or left-handedness of a helix does not depend on the direction from which it is observed. Follow the sense of rotation in the figure to check this. Normal DNA (B-DNA) forms a right-handed helix.

In other words, they cannot be interconverted by simple deformation, no matter how hard we try. If we imagine that the circle with four turns is completely relaxed, the circle with three turns will be underwound and the circle with five turns overwound. In order to attempt to return to the relaxed conformation (10.4 bp/turn) the double-helix will wrap over itself and become supercoiled as shown in Fig. 4.3. The molecule with a deficit of turns will form negative supercoils (the natural situation for DNA) and that with an excess of turns will form positive supercoils. It is important to realize that supercoiling is a consequence of DNA topology but is not a topological transition. It is easy to see how the supercoiled and non-supercoiled conformations can be interconverted by simple deformation. This is

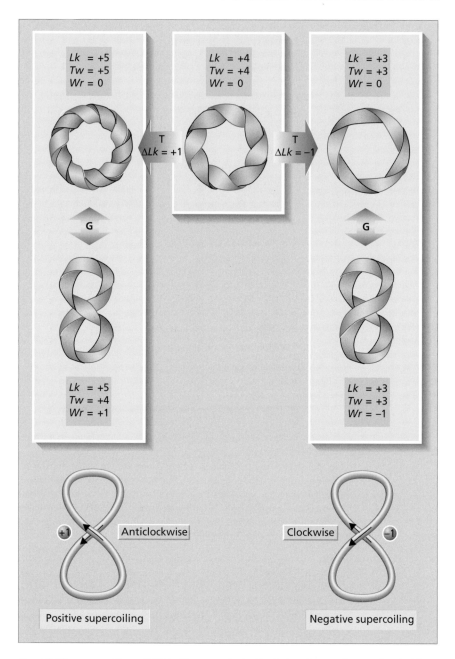

Fig. 4.3 The supercoiling of DNA. Three topologically distinct forms of a circular double-helix are represented. Imagine that the form with 4 turns of the helix happens to be totally relaxed. Because it has 4 helical turns it is described as having a linking-number of four. The form with 3 helical turns is stressed and can return to a relaxed helical pitch by forming half a super-helical turn. Similarly the form with 5 helical turns can return to a relaxed helical pitch by forming half a super-helical turn in the opposite direction. Each of the three topoisomers is defined by its linking number which remains constant. This can be partitioned into twist and writhe which are related by the equation

$$Lk = Tw = Wr$$

described in Fig. 4.3 which introduces the term **linking-number** and its partition into **twist** and **writhe**.

Knots are also topologically distinct forms of circular molecules. Two simple knots are shown in Fig. 4.4. and it is easy to see that they cannot be interconverted without breaking and joining the DNA. They are, therefore, topoisomers. If two DNA circles interwind, they can also form topoisomers and two of these are also illustrated in Fig. 4.4. Here we call the circles catenated and the structure a **catenane**.

The above examples show that there are two primary causes for topologically distinct forms of DNA. The first is the interwinding of 'Watson' and 'Crick' strands and the second is the interwinding of DNA double-strands in knots and catenanes. These are distinct but interconvertable 'currencies' of DNA structure. The interwinding of 'Watson' and 'Crick' strands is partitioned into twist and writhe; writhe can be converted into knots or catenanes by recombination. The total linkage of a molecule can therefore be expressed as a sum of twist, writhe, knotting and catenation as described in Fig. 4.4. By analysing the topology of the substrates and products of site-specific recombination reactions it has been possible to gain significant insight into their mechanisms.

4.3   ## Tn3 co-integrate resolution

The transposon Tn3 transposes by a replicative mechanism to generate a co-integrate structure, as described in Chapter 5. This co-integrate contains directly repeated copies of the transposon which can recombine with each other to generate two DNA circles, each carrying one copy (Fig. 4.5). This recombination is facilitated by an efficient site-specific recombination system encoded by the transposon. The *tnpR* gene encodes a protein, resolvase, which acts on an internal resolution site, *res*. DNA sequence

---

Fig. 4.3 (*Continued*) One full turn of the double-helix is defined as one unit of twist and can be precisely removed by supertwisting the DNA by half a turn. One half turn of the super-helix is therefore defined as one unit of writhe. This half turn appears as a crossing point (or **node**) when the DNA is projected onto a two-dimensional surface. You will notice that, if an arbitrary direction is given to the DNA double-strand, there are two types of node possible. These are defined as – or ± depending on whether the shortest possible rotation to map the top double-strand onto the bottom double-strand is clockwise or anticlockwise. Negatively supercoiled DNA has a deficit of helical turns which results in the formation of negative nodes of writhe. The partition between twist and writhe will depend on the thermodynamically most favoured conformation. The twist will want to approximate the relaxed pitch of the helix but there is a cost to significant amounts of writhe in terms of double-strand–double-strand repulsion which will mean that, in practice, twist will not return to exactly that favoured by the relaxed pitch. Note that, although the value of Lk must be integral and is a topological property of the molecule, the values of Tw and Wr need not be integral and are not topological properties. Topological transitions are marked with T and geometric or conformational changes are marked with G.

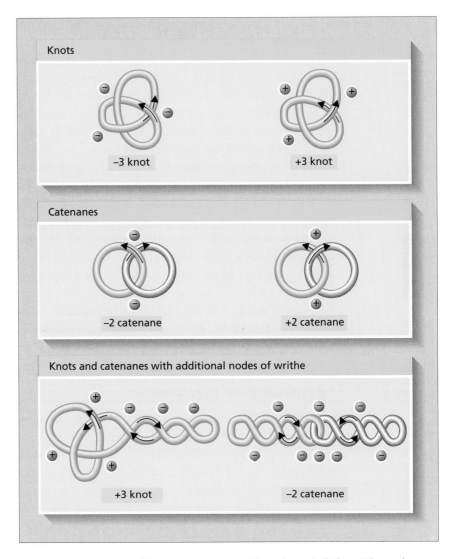

Fig. 4.4 Site-specific recombination is accompanied by a change in linkage. The total linkage of a molecule (*Lg*) can be described by the sum of its linking number (*Lk* = *Tw* + *Wr*) and its catenation (*Ca*) and/or knotting (*Kn*).

$$Lg = Lk + Ca + Kn \quad \text{or:}$$

$$Lg = Tw + Wr + Ca + Kn$$

Site-specific recombination will be accompanied by a change in total linkage (Lg) and/or its partition into *Lk*, *Ca* and *Kn*. In Fig. 4.3, we saw how the linking-number can be partitioned into twist and writhe and how writhe can be visualized as the number of crossings (or nodes) observed when the molecule is projected onto a two dimensional surface. Similarly, knots and catenanes can be visualized as having a number of nodes when the DNA is projected onto a two-dimensional surface. As with writhe, the nodes are positive or negative as defined by the rule in Fig. 4.3 and one node of writhe is equal to one node of knotting or catenation. However, since knotting and catenation are topological analysis has revealed that *res* has the complex structure shown in Fig. 4.5.

Randy Reed purified resolvase and showed that it catalysed site-specific recombination *in vitro*. The reaction generated catenated products which could be separated by restriction enzyme cleavage as shown in Fig. 4.6. This *in vitro* reaction has allowed several groups, notably those of Nick Cozzarelli, Martin Boocock and Dave Sherratt to study the topology of the substrates and products of the reaction. Wasserman and Cozzarelli demonstrated very elegantly that, not only were the products of recombination catenated, but that they were almost always a particular type of catenane (Fig. 4.7). This catenane was the simplest type possible, two singly interlinked circles. Furthermore, the two nodes of this catenane were always negative as defined by the convention described in Fig. 4.3. This demonstrated that the *res* sites must interact within a defined super-helical structure such as that shown in the currently accepted model of recombination described in Fig. 4.8.

The change in total linkage (Lg) associated with site-specific recombination is made up of the changes in linking number (Lk), knotting (Kn) and catenation (Ca) as described by the equation

$$\Delta Lg = \Delta Lk + \Delta Ca + \Delta Kn$$

for *res* mediated resolution we have seen that $\Delta Ca = -2$ and $\Delta Kn = 0$. Therefore, when ΔLk (which is simply the sum of the change in linking number of the two catenated circular products) was found to be $^{+}4$, the total change in linkage was calculated. We can see that ΔLg must have a value of $^{+}2$. This change in total linkage reflects the topological change that accompanies the breaking and joining of strands during recombination. The value of $^{+}2$ for ΔLg strongly suggests breakage followed by 180° rotation and rejoining of strands, as shown in Fig. 4.9 (a).

Fig. 4.4 (*Continued*) properties, nodes can only be added or removed in integral steps and *Kn* and *Ca* can only have integral values.

The simplest examples of knots and catenanes are depicted. The simplest knot has three nodes which can be either positive or negative and the simplest catenane has two nodes which can be either positive or negative. It is straightforward to determine the sign of a node in a knot. Simply assign an arbitrary direction to the DNA with an arrow and follow this round the molecule. This works well for knots but what about catenanes? Since there is no covalent continuity between the circles of a catenane, how can one ascribe a meaningful direction to each? The answer, in studies of site-specific recombination, is to define the direction of the arrow prior to recombination (as explained in Figs 4.7 and 4.8). These simplest forms of knots and catenanes can also have additional nodes of writhe as shown. The +3 knot with negative nodes of writhe is one of a number of structures that might arise from a λ site-specific inversion reaction and the −2 catenane with negative nodes of writhe is the structure generated by resolution by Tn3 resolvase.

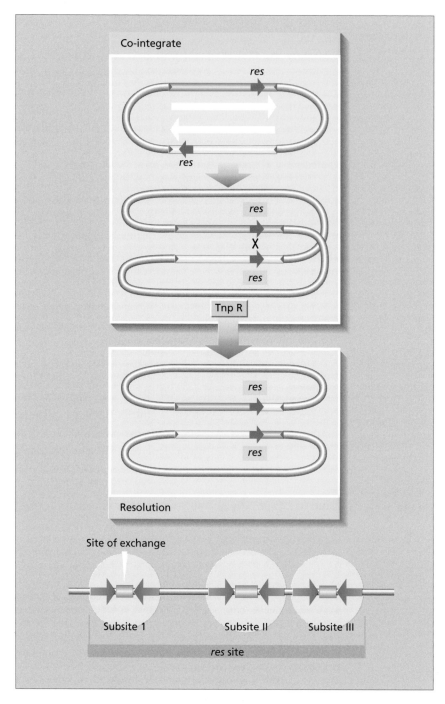

Fig. 4.5 Co-integrate resolution by Tn3 resolvase. Transposition of Tn3 generates a co-integrate structure which contains two directly repeated copies of the transposon. These are acted upon by the product of the *tnpR* gene, TnpR or resolvase, to generate the final products of transposition. Resolvase-mediated recombination occurs at the *res* site which is composed of subsites I, II and III. Each of the subsites consists of an imperfect inverted repeat of 9 bp, represented by an arrow, and a small spacer, represented by a box. The site of exchange is located at the centre of subsite I.

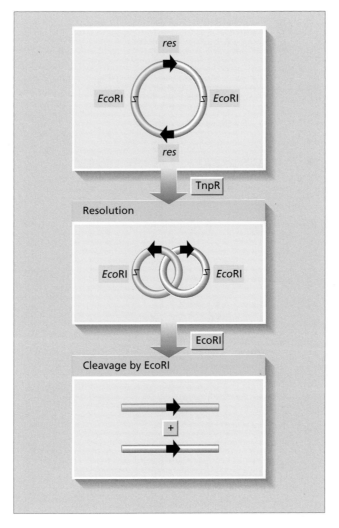

Fig. 4.6 *In vitro* **resolution with Tn3 resolvase.** Randy Reed (1981) showed that resolvase (TnpR) can mediate resolution of an artificial substrate with two directly repeated copies of the *res* site *in vitro*. This reaction requires only a buffer containing magnesium ions, a supercoiled substrate and a high molar ratio of resolvase to DNA. The products of the reaction are, however, catenated, and cleavage with a restriction enzyme such as *Eco*RI is required to separate the two product circles.

There are three possible reactions that can be catalysed by a site-specific recombination system, depending on the orientation of the sites and the nature of the substrates. These are: resolution, inversion and fusion, as depicted in Fig. 4.10. The fact that Tn3 resolvase can only catalyse resolution requires an explanation. Initially it was thought that the recombining sites might recognize each other by what has been called a tracking mechanism. That is, resolvase binds to one site and 'pulls' DNA past itself in an

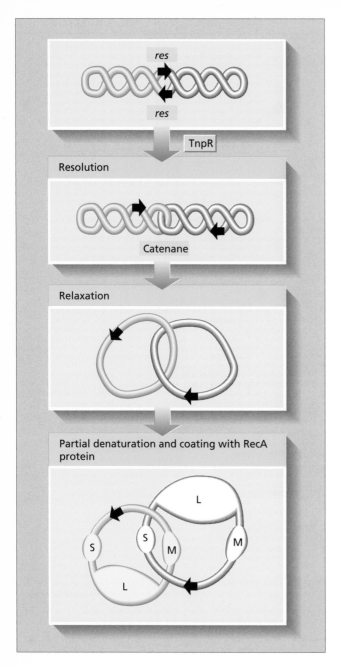

Fig. 4.7 The demonstration that resolution generates a ⁻2 catenane. After Wasserman & Cozarelli (1985). Resolution is shown here within a supercoiled DNA molecule. The product is therefore a two-noded catenane with extra nodes of writhe. Relaxation of the writhe within this structure generates two simply interlocked circles; but how can one determine the sign of the two nodes? To do this, two properties of the DNA must be known; firstly, which strand lies on top of the other and secondly, what is the orientation of each circle. These properties were both determined in ingenious ways. To establish which strand lies on top of the other, the DNA was coated with RecA protein before viewing under the electron microscope. This treatment makes the path of the DNA much wider and it is possible to see which strand passes over the other in electron micrographs. To determine the orientation for each of the circles, partial denaturation was used to establish the order of three loops of different sizes that melt before the rest of the circle. These regions lay in direct repeat in the DNA substrate and therefore defined the orientation of the DNA path in the product. The result was clear, only ⁻2 catenanes were formed.

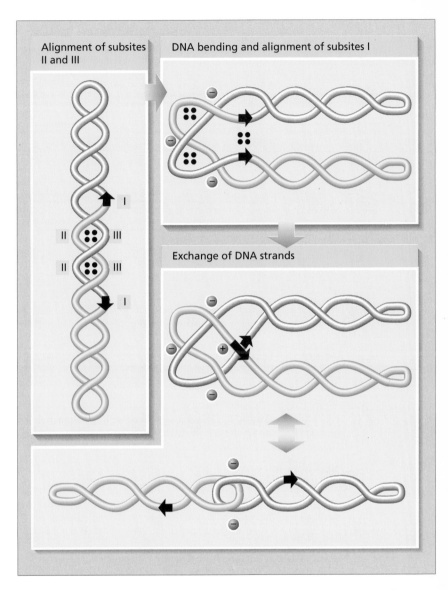

Fig. 4.8 Model for Tn3 Resolvase mediated recombination. Initially subsites II and III pair within supercoiled DNA. Resolvase (TnpR) is depicted as a black circle and a tetramer is believed to interact with each pair of subsites. Resolvase-binding causes DNA-bending and juxtaposition of subsites I where the exchange of strands takes place. This is mediated by another tetramer of resolvase. We can see that the synaptic complex traps three negative nodes of writhe between the recombining subsites and that recombination is accompanied by the formation of one positive node. When unfolded, this is the ⁻2 catenane observed for the reaction.

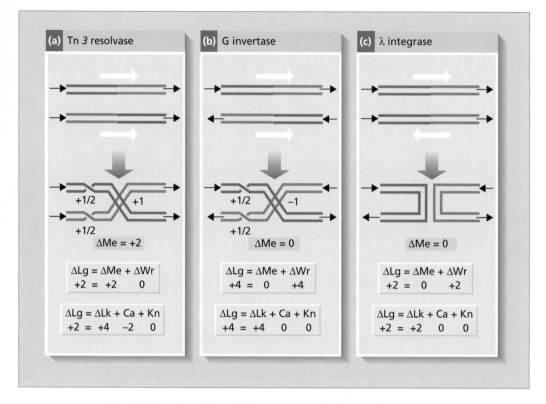

Fig. 4.9 The change in total linkage associated with recombination. Experiments have demonstrated that resolvase-mediated recombination results in a ⁻2 catenane ($\Delta Ca = {}^-2$) and a change in linking number of ⁺4 ($\Delta Lk = {}^+4$). From the equation

$$\Delta Lg = \Delta Lk + \Delta Ca + \Delta Kn$$

we can calculate that the change in total linkage associated with recombination must be ⁺2. This change in total linkage is a direct reflection of the topological change involved in the mechanism of the reaction. In other words, for directly repeated recombination sites

$$\Delta Lg = \Delta Me$$

where ΔMe is the topological change that occurs as a consequence of the strand-exchange mechanism. We can, therefore, imagine the DNA strands in the vicinity of subsites I isolated from the rest of the molecule and ask ourselves what mechanism of strand-exchange would be responsible for a ⁺2 change in linkage. In (a) we can see that cleavage of all four recombining strands accompanied by a right-handed (clockwise) rotation of 180° results in such a ⁺2 change. This is divided between a ⁺1 change in writhe (a ⁺1 node at the point of crossing over) and a ⁺1 change in twist (two ⁺1/2 turns of the helix must accompany the 180° rotation). The relative orientations of the strands, as defined by their connections, represented by small arrows and the orientation of the recombining sites is represented by large white arrows. For inversion reactions, ΔLg does not correspond directly with the topological change accompanying strand-exchange because nodes of writhe can also be lost or gained by virtue of the change in direction of the DNA path that must occur upon inversion. Thus:

$$\Delta Lg = \Delta Me + \Delta Wr$$

where ΔMe is the topological change associated with the mechanism of strand-exchange and ΔWr is the change in writhe caused by the trapping of nodes that change sign as a

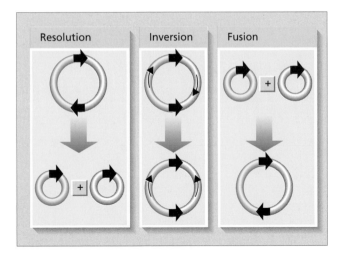

Fig. 4.10 **Three reactions catalysed by site-specific recombination systems.** Directly repeated sites result in resolution, inverted sites result in inversion and sites on separate molecules recombine to generate fusions.

oriented way until the second site is brought into contact with the complex. This hypothesis, and two experiments performed to test it, are illustrated in Fig. 4.11. This work demonstrated conclusively that the tracking hypothesis was incorrect and suggested instead that the orientation of the sites within super-helical DNA determined their ability to interact productively.

Trapping of intermediates has demonstrated that the reaction proceeds via a covalent intermediate in which the 5′ end of the cleaved DNA is joined to a serine residue of resolvase before strand-transfer is completed. Two pairs of strand-transfers are required to form the product of the reaction.

4.4   Bacteriophage λ integration and excision

Bacteriophage λ is a temperate phage. It has a choice; it can go through a lytic cycle, lysing its host cell and producing a burst of approximately 100

Fig. 4.9 (*Continued*) consequence of recombination. Also, because of the change in direction of the DNA path consequent upon inversion, nodes generated by the same mechanism for repeated and inverted sites will have opposite signs. It is believed that the same mechanism of recombination accompanies inversion of the G region by the Gin recombinase as does Tn*3* resolvase-mediated resolution (b). However, because of the change in the direction of the DNA path for the inversion reaction, the sign of the node created by crossing the strands is different (see also Fig. 4.18). For λ site-specific recombination (c) there is no net change in topological value accompanying the exchange of strands and for the simplified reaction studied by Nash & Pollock (1983) no catenation or knotting of the product. The λ integration reaction depicted here is the simplified 'integrative inversion' reaction studied by Nash and Pollock (see also Fig. 4.16).

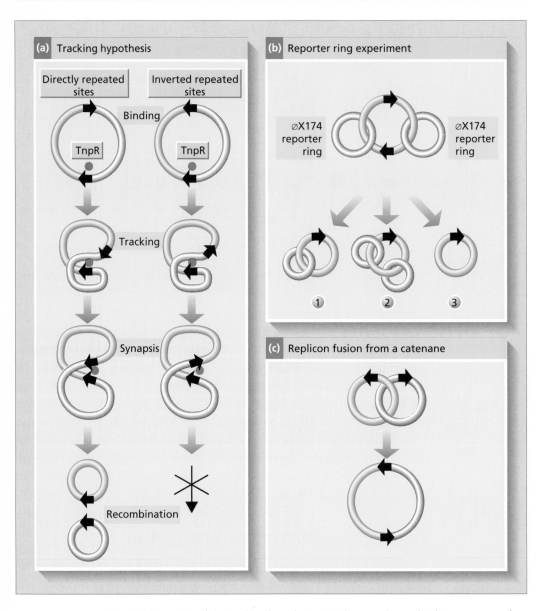

Fig. 4.11 Two tests of the tracking hypothesis. (a) The recombinase binds to one copy of the site and then actively tracks along the DNA while remaining bound at the first site. This brings the second site into conjunction and synapsis can occur. However, because of the oriented nature of tracking, the relative orientation of the sites in the synaptic complex is different for directly repeated and inverted sites. The hypothesis is that the relative orientation obtained for directly repeated sites is productive for Tn3 resolution but that for inverted sites is not productive for inversion. Two predictions can therefore be made if tracking is correct. Firstly, the two loops generated by tracking are not equal. One (blue in this figure) starts very small and grows while the other (red in this figure) starts large and gets smaller. The loops are however closed and any catenated rings should not be able to pass from one loop to the other. Secondly, tracking explains the orientation effect by

particles, but can also become dormant by repressing lytic gene expression and integrating into the host chromosome. The dormant λ genome, or prophage, was shown to have a different gene order to that of the lytically grown phage (Fig. 4.12). This led Alan Campbell (1961), to propose the revolutionary idea that the λ genome circularized and formed a prophage by recombination with its host. He suggested that this recombination

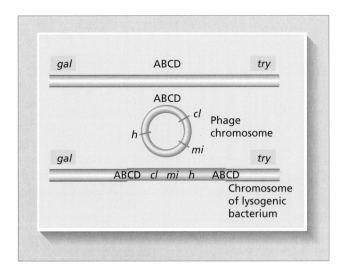

Fig. 4.12 Permutation of the λ gene order in the prophage. After Campbell, 1961. In 1960 (Calef & Licciardello) it was demonstrated that the gene order of a λ prophage was different to that of a vegetatively growing phage. The prophage map was *try-h-mi-cl-gal* whereas the vegetative map was *h-cl-mi*. This prompted Alan Campbell (1961) to propose that the phage genome might circularize and integrate into the host chromosome by recombination. The site denoted ABCD is what we now know of as the attachment site (*att*).

Fig. 4.11 (*Continued*) postulating that the DNA is passed in front of the bound site in a direction that is determined by the orientation of the first site. Tracking, therefore, cannot occur after random collision between DNA molecules and requires covalent continuity between the first and the second site. These two predictions have been tested and both have been shown to be incorrect. (b) The reporter ring experiment used two rings of Φ×174 linked to a substrate for resolution (Benjamin, Matsuk, Krasnow & Cozzarelli, 1985). The prediction of the tracking hypothesis is that very few molecules of type 1 should occur. Tracking would start with a small loop unlikely to contain a reporter ring which would grow but would not gain a ring during its growth. Products of type 2 and 3 would therefore greatly predominate. In fact, 40% of the products were of type 1, arguing for a mechanism where sites can interact, trapping the reporter rings in separate loops. (c) Since the tracking hypothesis requires covalent continuity between the two recombining sites, two catenated circles should not recombine. However, it was shown that catenated circles could recombine. (Stark, Sherratt & Boocock, 1989). This occurred only at low super-helical densities because the change in linking number associated with this reaction was ‑4, a change that would not be favoured by negative super-helicity. Nevertheless, the observation that this reaction was possible argued against a tracking mechanism.

occurred at an internal site, resulting in the permuted gene order that had been observed.

We now know that Campbell's hypothesis was correct, and that λ encodes two gene products (Int and Xis) that mediate integration and excision. These proteins act in conjunction with host proteins IHF, and FIS to mediate site-specific recombination between the phage attachment site *att*P (POP') and the bacterial attachment site *att*B (BoB') as shown in Fig. 4.13. Int

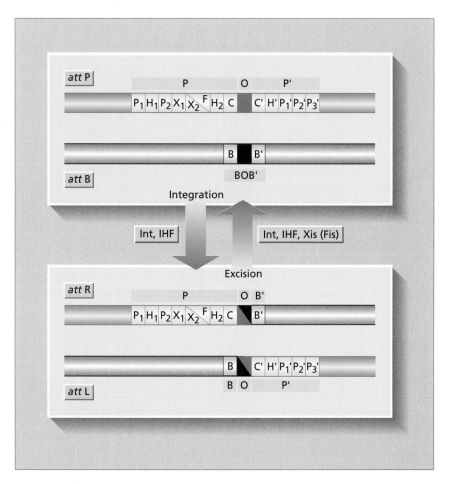

Fig. 4.13 Integration and excision of bacteriophage λ. The phage attachment site (POP' or *att*P) is very complex and contains binding sites for Int (P_1, P_2, C, C', P'_1, P'_2, P'_3), Xis (X_1, X_2), IHF (H_1, H_2, H') and Fis (F). The Int binding sites are of two types, arm-type (P_1, P_2, P'_1, P'_2, P'_3) and core type (C, C'). The protein makes bridge-like contacts between the arm and core-type binding sites with the N-terminal part of Int bound to the arm-type sites and the C-terminal part bound to the core-type sites. The bacterial attachment site (BOB' or *att*B) is much simpler with only core-type Int binding sites. Both POP' and BOB' contain the overlap region (O) that lies between the sites of strand-transfer. Recombination generates the two hybrid sites (BOP', or *att*L, and POB' or *att*R).

and IHF are required for integration, while for excision an additional protein, Xis, is also needed. The protein Fis, which also enhances the unrelated inversion reactions described in Sections 4.5 and 4.6 below, is required at sub-optimal concentrations of Xis such as those found *in vivo*. It is interesting to note that the regulation of *fis* gene expression is controlled by the growth rate of the cell. In log-phase cells, Fis is made, whereas in stationary-phase cells it is not. This makes sense, since λ should only excise in a log-phase cell capable of supporting lytic growth of the phage. Ingenious experiments carried out by the groups of Art Landy and Howard Nash have demonstrated that integration and excision proceed by two pairs of strand-transfer reactions. The first occurs at the left-hand side of the 7 bp homologous overlap region of the *att* sites to form a 4-way junction. This junction then migrates to the right side of the overlap and a second pair of strand-transfers completes the reaction (Fig. 4.14). As for Tn3 resolvase, the reaction proceeds via a covalent intermediate between the DNA and the protein. However, here, the 3′ phosphate of the cleaved DNA is joined to a tyrosine residue of integrase before strand-transfer is completed.

The topological analysis of λ integration and excision has been more difficult than that of Tn3 co-integrate resolution. This is because of three complexities. Firstly, the λ site-specific recombination reactions allow inversion, resolution and fusion depending on the orientation and disposition of the sites. Secondly, the recombining sites are non-identical, which leads to a large number of possible forward and reverse reactions if one takes into account inversion, resolution and fusion for each pair of sites. Thirdly, the nature of the catenated or knotted products of the reaction depend on the trapping of random numbers of nodes of writhe present in the substrate as well as any nodes of writhe trapped by the mechanism. Some of the products of λ site-specific recombination are shown in Fig. 4.15. What is established by these reactions is that the two recombining sites can form a productive synapse after random collision with each other. This contrasts sharply with the situation observed for Tn3 resolvase where the recombining sites must interact within in a precise structure in supercoiled DNA. This difference in mechanism is what accounts for the difference between the restricted behaviour of Tn3 resolution and the relaxed behaviour of λ site-specific recombination with respect to the orientation and disposition of the sites. Despite the experimental difficulties, Nash and Pollock have shown that an inversion reaction between POP′ and BOB′ on a partially relaxed substrate results in an unknotted product with a +2 change in linking number ($\Delta Lk = +2$). This is consistent with several geometries, one of which is depicted in Fig. 4.16, and a strand-exchange topology (depicted in Fig. 4.9 (c)).

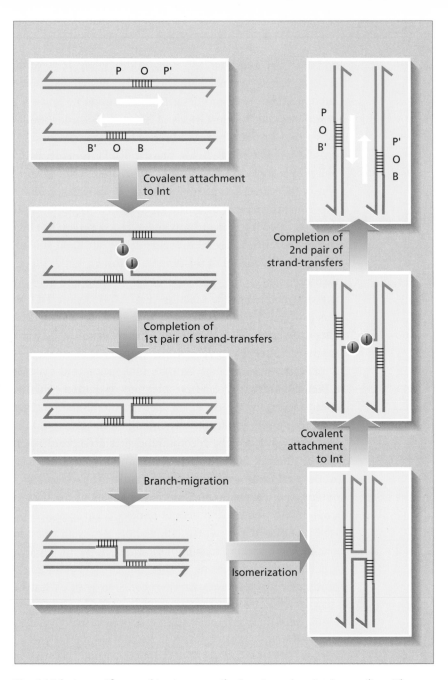

Fig. 4.14 λ site-specific recombination proceeds via a 4-way junction intermediate. The phage (POP') and bacterial (BOB') attachment sites are shown aligned in an anti-parallel orientation with respect to each other (see the large arrows). The 'O' or overlap region is depicted as 7 bp connecting each of the two strands. Initially, Int protein cleaves the left-hand side of the overlap and forms a covalent link to the 3′ phosphate generated. The first pair of strand-transfers then takes place and a 4-way junction is formed. This structure can branch migrate and isomerize to bring the junction into a configuration where the second pair of parental strands can be cleaved and strand-transfer completed via covalent DNA-Int intermediates at the right-hand side of the overlap region.

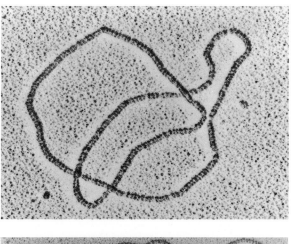

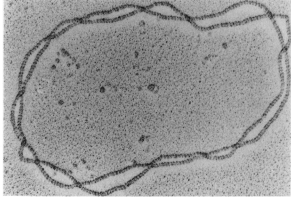

Fig. 4.15 Some products of λ site-specific recombination, as in Spengler, Stasiak & Cozzarelli (1985). Electron micrographs of two products of Int-mediated recombination are shown. Both are derived from a substrate with inverted recombination sites. The first is a knot with 3 positive nodes and the second is a knot with 13 positive nodes. The path of the DNA can clearly be visualized because of the coat of RecA protein that has been applied. Int-mediated inversion results in the formation of a wide variety of different knotted structures that arise due to the trapping of random numbers of nodes of writhe between the recombining sites. Photomicrographs kindly provided by N. Cozzarelli (University of California, Berkely).

4.5

Bacteriophage host-range variation

Bacteriophage Mu and several other phages, including P1, have evolved a mechanism to increase their host-range. They do this by encoding two alternative tail fibres which allow adsorbtion to different cell surface lipopolysaccharide receptors. This involves the inversion of a DNA segment to alternate gene expression. Mu has two tail fibre genes, S and U. The S gene has a constant part (S_c), that lies outside of an invertable region and

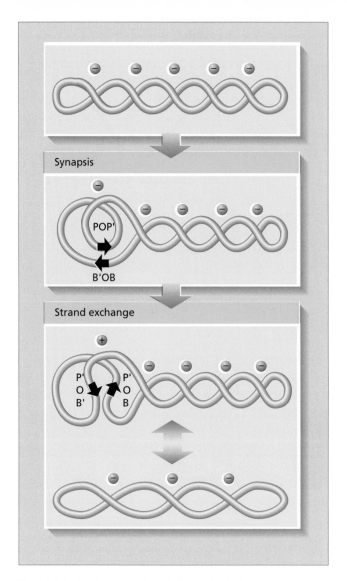

Synapsis

POP'

B'OB

Strand exchange

P'
O
B'

P'
O
B

Fig. 4.16 A simple explanation of the topological change accompanying λ integration. In an experimental system for looking at λ integration, (Nash & Pollock, 1983) the phage and bacterial attachment sites were introduced into one DNA molecule and oriented in inverted repeat so that recombination inverted the DNA between the sites. The sites were located close to each other (570 bp apart) and the substrate was relaxed to about $\frac{1}{4}$ of its normal super-helical density to avoid the accidental trapping of super-helical nodes that happen to lie between the sites. This procedure reduced the number of knotted products so much that a significant proportion of unknotted molecules were produced. Their linking-number was compared to that of the substrate and the change determined as $\Delta Lk = {}^+2$. In this explanation of the result, the sites line up in anti-parallel orientation as suggested previously (Stark, Sharratt & Boocock, 1989) and a negative node is changed into a positive node as a consequence of recombination. This constitutes the ${}^+2$ change in linking-number and implies that no other contributions to the linking number change are made by the mechanism of recombination. This is shown in Fig. 4.9 (c), where we can see that the exchange itself is 'zero-rated' as opposed to the ${}^+2$-rating of resolvase-mediated recombination.

a variable part (S_v) within it. As shown in Fig. 4.17, inversion of this DNA segment (the G region) causes alternate expression of S_cS_v and $S_cS'_v$ and of U and U'. It is interesting to note that inversion is slow relative to the lytic cycle of Mu. Therefore, when Mu is growing lytically, little inversion is observed and almost all the phage have the tail fibres appropriate for the host that is being infected. Mu is, however, a temperate phage and can exist as a prophage. When this happens, inversion has time to go to completion and 50% G⁺, 50% G⁻ prophages exist within a population. Therefore, when a prophage induces, the virions produced benefit from having the extended host range this system provides.

Inversion of the G region is catalysed by the protein Gin which is the product of the *gin* gene located just outside of the G region on the Mu genome. Inversion is stimulated by an *Escherichia coli* protein called FIS which we have also seen stimulates λ excision. Here, FIS binds to a recombinational enhancer sequence which acts in a distance- and orientation-independent manner. Gin is a distant relative of Tn*3* resolvase and shares its restricted mode of action. However, here, Gin catalyses inversion but not resolution. This is because Gin, like resolvase, requires a very specific

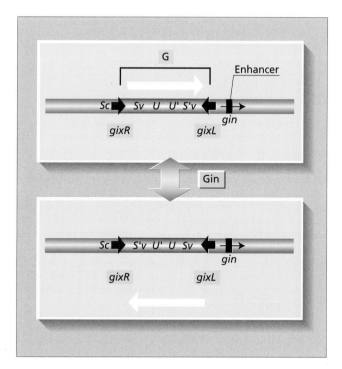

Fig. 4.17 Inversion of the G region of bacteriophage Mu. The Gin protein mediates inversion of a 3 kb region of DNA that is responsible for encoding the tail fibres of the phage. Inversion occurs at the Gix sites denoted by filled arrows. The *gin* gene itself is located outside the invertable region.

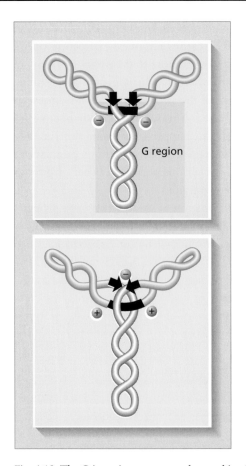

Fig. 4.18 The G inversion synapse and recombination. It is proposed that two negative nodes are trapped in the G synapse between the *Gix* sites (arrows) and the enhancer sequence (box). Recombination converts these negative nodes into positive nodes and one negative node is generated by the crossing of the strands. This is a total linkage change of ⁺3 and leaves an unknotted product. Remember however that the observed change in linking number ΔLk is ⁺4. Since there is no knotting or catenation in the reaction $\Delta Lg = \Delta Lk$ and there must be another ⁺1 to bring the total linkage change to ⁺4. This, as for resolvase, is a twist change required for the 180° rotation. This twist change will be partitioned between twist and writhe but is represented diagrammatically here as a change in writhe. The close-up of the exchange of strands (Fig. 4.9 (b)) shows that the same mechanism can account for Tn3 resolution and G inversion and the difference in sign between the node generated by the crossing of the DNA strands (+1 for resolution and −1 for inversion) is because of the inversion of the DNA path consequent upon recombination, not to the recombination reaction itself.

synapse within supercoiled DNA (Fig. 4.18). A change of ⁺4 in linking number ($\Delta Lk = {}^{+}4$) has been determined for this reaction consistent with a mechanism using a right-handed 180° rotation like that used by Tn3 resolvase (see Fig. 4.9 (b)).

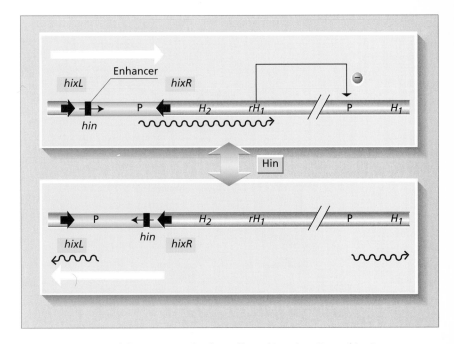

Fig. 4.19 Inversion of the H region of *Salmonella typhimurium*. Recombination occurs between the *hix* sites (arrows) to invert the H region. In contrast to G inversion, the Hin recombinase and the recombinational enhancer are encoded within the invertable region. Recombination inverts the promoter of the *H2* gene that encodes H2-type flagellae. The same promoter is used by the *rH1* gene which is a repressor of gene *H1* expression. When inverted, *H2* and *rH1* are not expressed and this allows expression of *H1* and synthesis of H1-type flagellae.

4.6 ## Antigenic variation

Another system related to G inversion and Tn3 co-integrate resolution is flagellar variation in *Salmonella*. *S. typhimurium* encodes two types of flagella which alternate due to inversion of a DNA segment called H. In this case, the recombinase is actually located within the invertable region, as is the recombinational enhancer (Fig. 4.19).

4.7 ## DNA replication

The yeast *Saccharomyces cereviciae* normally maintains a small multi-copy plasmid called 2 μm. This plasmid is not believed to confer any advantage to its host and may best be considered a molecular parasite. It replicates efficiently and uses DNA inversion to increase its copy number, as shown in Fig. 4.20. The invertable region (approximately 40% of the molecule) is bounded by inverted repeats of a site, FRT, that is acted upon by the recombinase FLP. FLP is distantly related to the Int protein of bacteriophage λ

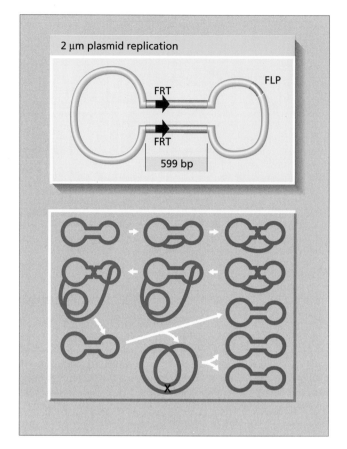

Fig. 4.20 Copy number amplification by the action of FLP recombinase. After Cox, 1988. The FRT site lies within 599 bp inverted repeats on the 2 μm plasmid. Futcher proposed that recombination allows the formation of a double rolling-circle molecule that can produce many copies of the plasmid from one replication initiation event.

and the mechanism of recombination proceeds via a Holliday intermediate like the λ reaction. In contrast to λ, however, the two recombining sites are identical and no protein other than FLP is required for the reaction.

4.8   Chromosome monomerization

The chromosome of *E. coli* has a system to ensure that dimers are efficiently resolved to monomers. This is important because a dimeric chromosome can only be inherited by one daughter cell at division. This would result in the formation of chromosome-free cells. Dimers of the chromosome can occur as a consequence of sister-strand exchange, as shown in Fig. 4.21. A site, denoted *dif*, ensures that these dimers do not persist. *Dif* is acted upon by two gene products XerC and XerD. Both of these are recombinases that belong to the Integrase family of proteins.

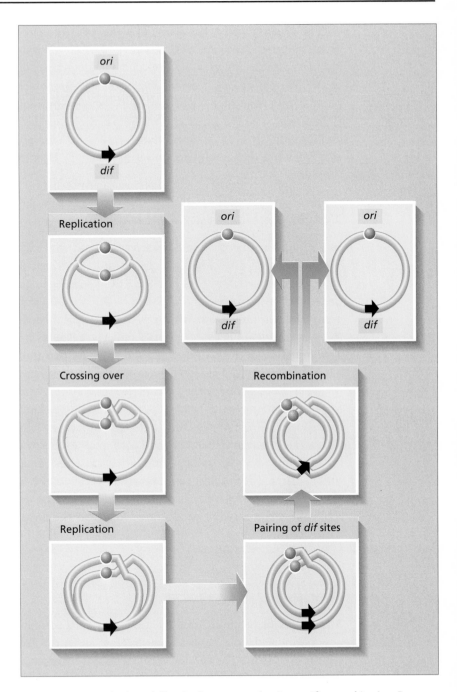

Fig. 4.21 Monomerization of dimeric chromosomes by site-specific recombination. Because bacterial chromosomes are circular, homologous recombination events between fully or partially replicated sister molecules can result in dimeric structures. It is advantageous that these should be monomerized efficiently to prevent chromosome-free cells from forming at cell division. *E. coli* has a site *dif* which is acted upon by the products of the *xerC* and *xerD* genes to efficiently resolve these dimers back to monomers.

The plasmid ColE1 has acquired a site analogous to *dif* that is known as *cer* and uses the same gene products to catalyse monomerization. This also performs the function of promoting the inheritance of at least one copy of the plasmid at cell division. Bacteriophage P1, which can reside as a plasmid prophage of *E. coli*, encodes its own site-specific recombination system related to the Integrase family that performs the same purpose. In this case, the site (*lox*) is acted upon by a P1 encoded protein (Cre) which also facilitates the circularization of the phage DNA after injection into a new host.

4.9   Mobile gene-cassettes and integrons

Several operons encoding resistance to antibiotics have been found that are composed of **cassettes** which can be integrated and excised by site-specific recombination as shown in Fig. 4.22. The integrases (IntI) that mediate these reactions are related to λ integrase and are located just upstream of a primary attachment site, *attI*. A promoter reads through *attI* and is used to transcribe genes inserted into the site. The cassettes normally consist of a promoterless gene and a downstream sequence of approximately 50–100 bases denoted a '59-base element' after the first of these to be character-ized. The 59-base element is an imperfect palindrome that contains the site for recombination with *attI*. Operons containing several genes can be formed by insertion of cassettes at *attI* sites adjacent to integrated cassettes. The whole structure, including *intI*, *attI* and any number of cassettes, is called an **integron**. It has been found that cassettes can also be deleted by IntI-mediated site-specific recombination between the 59-base elements. Furthermore, the 59-base elements themselves can be deleted to generate stable junctions between genes. It is likely that the imperfect palindromic nature of the 59-base elements facilitates their deletion by strand-slippage, as described in Chapter 6.

4.10   Conclusion

All site-specific recombination reactions that have been investigated in both prokaryotic and eukaryotic systems fall into either the Tn3 resolvase or the λ integrase family. The former is characterized by recombination reactions that occur within highly defined geometries and result in precisely defined product topologies. This characteristic means that productive and non-productive orientations of sites exist. In contrast, the λ integrase family has a more relaxed requirement for the geometries of its recombination complexes and this results in more varied product topologies. This permits members of this family to catalyse resolution, inversion and fusion reac-

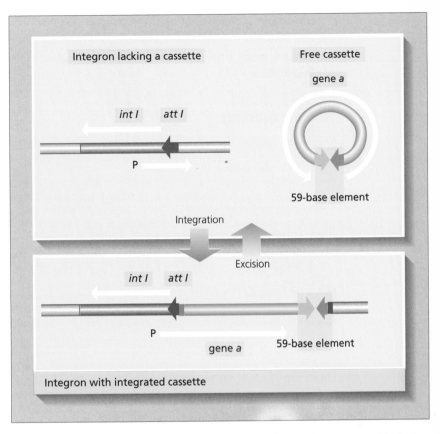

Fig. 4.22 Mobile gene-cassettes and integrons. Gene-cassettes can integrate at *attI* by IntI-mediated recombination. Circular cassettes have been detected and shown to integrate by recombination between the 59-base element and *attI*. The two arms of the palindrome forming the 59-base element are not identical and have therefore been represented by arrows of different shading. The *attI* sequence is even more different and is therefore represented by a different coloured arrow. Nevertheless, the external bases of both arms of the 59-base element and *attI* contain the sequence GTTRRRY (where R represents a purine and Y represents a pyrimidine) and recombinant junctions show that the exchanges have occurred between the G and the first T. Although IntI is related to λ integrase and IntI-mediated recombination is likely to be related to λ Int-mediated recombination, the fact that the recombining sites only share limited homology suggests that some details of the strand-transfer reactions must be significantly different.

tions depending on the orientation and disposition of the recombining sites.

Site-specific recombination is a precise and reciprocal mechanism and has, therefore, been utilized to catalyse recombination reactions where there is a biological requirement for the precise recovery of all the products. As defined in Chapter 1, such reciprocal events must proceed via an even number of pairs of strand-transfer reactions and here four strand-transfers occur in each case. They proceed via energetically neutral transesterification steps, initially to form DNA-protein intermediates and then to generate the exchanged strands. Despite the fact that the strand-transfers are energetically neutral, the reactions are favoured by negative supercoiling and normally proceed in the direction that results in some relaxation of super-helicity.

Further reading

Cox M. M. (1988). FLP site-specific recombination system of *Saccharomyces cerevisiae*. In Kulcherapati R. & Smith G. R. (eds) *Genetic Recombination*, pp. 429–443. American Society for Microbiology, Washington.

Cozzarelli N. R., Boles T. C. & White J. H. (1990). Primer on the topology and geometry of DNA supercoiling. In Cozzarelli N. R. & Wang J. (eds) *DNA Topology and its Biological Effects*, pp. 139–184. Cold Spring Harbor Press, New York.

Dröge P. & Cozzarelli N. R. (1992). Topological structure of DNA knots and catenanes. *Methods in Enzymology* **212**, 120–130.

Glasgow A. C., Hughes K. T. & Simon M. I. (1989). Bacterial DNA inversion systems. In Berg D. E. & Howe M. M. (eds) *Mobile DNA*, pp. 637–659. American Society for Microbiology, Washington.

Hall R. M. & Collis C. M. (1995). Mobile gene cassettes and integrons: capture and spread of genes by site-specific recombination. *Molec. Microbiol.* **15**, 593–600.

Hatfull G. F. & Grindley N. D. F. (1988). Resolvases and DNA-invertases: a family of enzymes active in site-specific recombination. In Kulcherapati R. & Smith G. R. (eds) *Genetic Recombination*, pp. 357–396. American Society for Microbiology, Washington.

Landy A. (1989). Dynamic, structural, and regulatory aspects of λ site-specific recombination. *Ann. Rev. Biochem.* **58**, 913–949.

Stark W. M., Boocock M. R. & Sherratt D. J. (1989). Site-specific recombination by Tn3 resolvase. *Trends Genet.* **5**, 304–309.

Stark W. M., Boocock M. R. & Sherratt D. J. (1992). Catalysis by site-specific recombinases. *Trends Genet.* **8**, 320–321.

Stark W. M. & Boocock M. R. (1995). Topological selectivity in site-specific recombination. In Sherratt D. (ed.) *Mobile Genetic Elements: Frontiers in Molecular Biology*, pp. 101–129. Oxford University Press, Oxford.

References

Benjamin H. W., Matzuk M. M., Krasnow M. A. & Cozzarelli N. R. (1985). Recombination site selection by the Tn3 resolvase: topological tests of a tracking mechanism. *Cell* **40**, 147–158.

Calef E. & Licciardello G. (1960). Recombination experiments on prophage host relationships. *Virology* **12**, 81–103.

Campbell A. M. (1961). Episomes. *Adv. Genet.* **11**, 101–145.

Cox M. M. (1988). FLP site-specific recombination system of *Saccharomyces cerevisiae*. In Kulcherapati R. & Smith G. R (eds) *Genetic Recombination*, pp. 429–443. American Society for Microbiology, Washington.

Nash H. A. & Pollock T. J. (1983). Site-specific recombination of bacteriophage lambda. The change in topological linking number associated with exchange of DNA strands. *J. Molec. Biol.* **170**, 19–38.

Reed R. R. (1981). Transposon-mediated site-specific recombination: a defined *in vitro* system. *Cell* **25**, 713–719.

Spengler S. J., Stasiak A. J. & Cozzarelli N. R. (1985). The stereostructure of knots and catenanes produced by phage λ integrative recombination: Implications for mechanism and

DNA structure. *Cell* **42**, 325–334.

Stark W. M., Sherratt D. J. & Boocock M. R. (1989). Site-specific recombination by Tn3 resolvase: topological changes in the forward and reverse reactions. *Cell* **58**, 779–790.

Wasserman S. A. & Cozzarelli N. R. (1985). Determination of the stereo structure of the product of Tn3 resolvase by a general method. *Proc. Natl. Acad. Sci. USA* **82**, 1079–1083.

5 Transposition

5.1 Introduction

Transposable elements were discovered because of the mutations they caused. In the 1940s and 1950s Barbara McClintock studied the effects of what she called '**controlling elements**' in maize and concluded that the insertion and excision of DNA sequences must be responsible for the patterns of pigmentation observed in kernels. However, it was not until the 1960s that similar processes were observed in prokaryotes, and with this discovery came the isolation and characterization of the DNA sequences involved.

In 1963, Larry Taylor described the first prokaryotic element—a bacteriophage that he called Mu because it caused *mu*tations. Like λ, Mu could lysogenize its host *Escherichia coli* but, unlike λ, it did not have a specific attachment site in the chromosome where its DNA integrated to form a prophage. Instead, Mu DNA integrated approximately randomly causing mutations wherever it disrupted a gene or regulatory site. An early observation was that approximately 1–2% of Mu lysogens had acquired a new auxotrophy. This frequency corresponds to that of auxotrophic loci on the chromosome.

In the late 1960s, a class of unusual, strongly polar, mutations of *E. coli* was discovered. These mutations were unusual in that they were not suppressed by the known suppressors of nonsense or frame-shift mutations but were not deletions because they could revert at a low but clearly detectable frequency. They were in fact caused by the insertion of small DNA sequences of approximately 1 kb in length that have been called **insertion sequences** or '**IS**' for short.

In the early 1970s, yet another group of prokaryotic elements was discovered; these were the **transposons**, so named because they conferred properties (e.g. drug resistance) that could be 'transposed' from one lo-

112

cation to another. The transposons were primarily associated with bacterial plasmids and have contributed to the rapid dispersal of resistance to clinically used antibiotics.

The study of the molecular nature of eukaryotic transposable elements was prompted by the discoveries in prokaryotes. Soon, a large number of different elements were characterized and grouped into four classes; the **transposons** that transpose via a DNA intermediate; the **retrotransposons** that use an RNA intermediate and are related to **retroviruses**; the **retroposons** that also use an RNA intermediate but do not share the structural features associated with retroviruses; and the **fold-back** elements that have an unusual repeated and palindromic structure.

Transposable elements are clearly a diverse family but one whose members have certain features in common. They are DNA sequences capable of insertion at several different chromosomal locations; they all have defined ends; and have non-permuted structures. They must also regulate their transposition tightly to prevent excessive mutagenesis.

5.2   Classification and structure of prokaryotic transposable elements

Nancy Kleckner proposed a useful subdivision of prokaryotic transposable elements into four classes. **Class I** elements include the insertion sequences (IS) and the **composite transposons** built from insertion sequences. Insertion sequences are small (~800–1500 bp), comprise inverted terminal repeats and at least one open reading-frame encoding a protein required for transposition (the **transposase**) but do not encode other functions (Fig. 5.1 (a)). Composite class I elements are formed by two insertion sequences co-operating to transpose not only their own DNA but also the DNA lying between them (see Fig. 5.1 (b)). This DNA can encode anything and many of these elements confer resistance to an antibiotic drug.

Class II elements are the transposons related to Tn3. In contrast to the composite transposons, these encode both transposition functions and other properties between the inverted repeats of the elements themselves. Tn3 encodes a transposase, a site-specific recombination protein (resolvase) and a β-lactamase responsible for ampicillin resistance (see Fig. 5.1 (c)).

Class III elements are the transposing bacteriophages, Mu and D108. These phages are closely related to each other and the structure of Mu is shown in Fig. 5.1 (d).

Class IV elements are those not easily placed in the other groups. The class may require further subdivision as more becomes known about its members. A well studied member is Tn7 which has five transposition genes

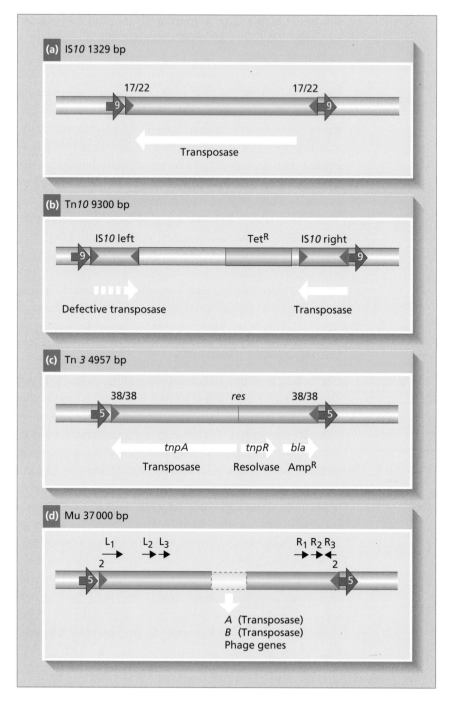

Fig. 5.1 The structure of prokaryotic transposable elements. Examples of prokaryotic transposable elements are shown. Terminal repeats are denoted by triangular arrowheads and target-site duplications by filled-in arrows. Genes are represented by white arrows below the elements. (a) **Class I: IS10.** The element has one large open reading-frame that

and two types of target-site: a major site that is used almost exclusively if it is present and minor sites that are used in its absence.

5.3

Classification and structure of eukaryotic transposable elements

The eukaryotic transposable elements can also be subdivided into four classes. Here though, there is the added complication that there are often both **active** and **defective** elements; the latter requiring the presence of an active element to provide gene products *in trans* to activate transposition. Defective elements that can be complemented for transposition retain the ends of the element that are recognized by the transposase but have internal deletions eliminating the transposase itself.

The first class are the **transposons** that transpose via a DNA intermediate. These include Ac (and its defective derivatives, Ds) from maize and the P factor (and its defective derivatives, P elements) from *Drosophila*. The structure and transposition of these elements resembles that of the prokaryotic transposable elements. The structure of Ac is shown in Fig. 5.2 (a). A subclass of eukaryotic transposons whose members are widely distributed in many different species from fungi to vertebrates is the Tc1/Mariner family. In this subclass are Tc1 and Tc3 from the nematode worm *Caenorhabditis elegans* and Mariner from *Drosophila*.

The second class are the **retrotransposons**. These are elements that resemble **retroviruses** in their genetic organization, transposition mechanism and nature of gene products. There are many characterized retrotransposons such as the Ty elements from yeast and Copia from *Drosophila*. The structure of Ty1 is shown in Fig. 5.2 (b).

The third class are the **retroposons**. Like the retrotransposons, these elements are believed to transpose via an RNA intermediate. Their genetic

Fig. 5.1 (*Continued*) encodes the transposase. It is bounded by imperfect inverted repeats of 22 bp in length. 17 out of the 22 bp in the inverted repeats are identical. IS*10* generates a 9 bp target-site duplication upon transposition. (b) **Composite Class I: Tn*10*.** Tn*10* is composed of two copies of IS*10* that flank DNA sequences encoding resistance to the antibiotic tetracycline (Tet^R). Only one copy of Tn*10* encodes an active transposase. The other copy carries mutations that inactivate transposase. (c) **Class II: Tn*3*.** Tn*3* has three principle open reading-frames. The *tnpA* gene encodes the transposase; *tnpR*, a site-specific resolvase; and *bla*, β-lactamase that is responsible for ampicillin resistance. The element is bounded by perfect 38 bp inverted repeats and generates a 5 bp target-site duplication upon transposition. (d) **Class III: Mu.** The ends of bacteriophage Mu are represented. Mu is both a transposable element and a bacteriophage. It encodes two transposition proteins, A and B. A is essential for transposition and binds to 30 bp sites L1–L3 at the left end and R1–R3 at the right end. These six binding sites share partial sequence identity. At the ends of the element there are inverted repeats of only 2 bp. Transposition results in a target-site duplication of 5 bp.

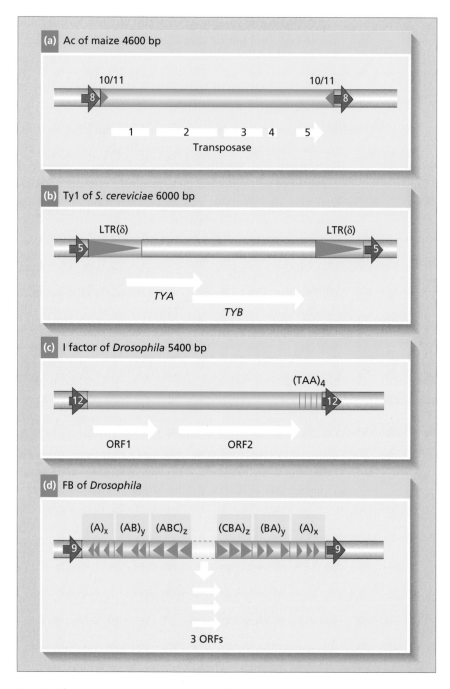

Fig. 5.2 The structure of eukaryotic transposable elements. Terminal repeats are denoted by triangular arrowheads and target-site duplications by filled-in arrows. Genes are represented by white arrows below the element. (a) **Transposons: Ac of maize.** This transposon resembles bacterial elements in that it transposes without an RNA intermediate. It has imperfect terminal inverted repeats and generates an 8 bp target-site duplication. The transposase is encoded by a single transcript that is composed of five exons (labelled 1–5)

organization, however, does not resemble the retroviruses. This class includes the mammalian LINE elements and the I factor of *Drosophila* which is shown in Fig. 5.2 (c).

The fourth class are the **fold-back** or **FB** elements. The FB elements are unusual in that they have internally repetitive long inverted repeats. Between the inverted repeats there is often DNA with open reading-frames that may encode transposition functions. The mechanism of transposition of FB elements is unknown but is not thought to involve an RNA intermediate. At present, FB elements have only been found in *Drosophila* and the nematode *C. elegans*. The structural organization of the *Drosophila* elements is shown in Fig. 5.2 (d).

5.4   ## Target-site duplication

Nigel Grindley sequenced a number of insertions of IS*1* in *E. coli* and discovered that the sequence that had originally been present at the site of insertion was duplicated so that it flanked the inserted element. This led to the proposal that the duplication arose as a consequence of the mechanism of insertion which involved cleavage of the DNA with **staggered breaks,** as shown in Fig. 5.3. Subsequently, when the insertion sites of other elements were sequenced, two common lengths of **target-site duplication** were found to be 5 and 9 bp. It is interesting to note that these lengths correspond

---

Fig. 5.2 (*Continued*) and four introns. (b) **Retrotransposons: Ty1 of *S. cerevisiae*.** Retrotransposons resemble retroviruses. Ty1 consists of two long terminal repeats (LTRs) that are also known as δ (331–338 bp in length). These flank two genes, *TYA* and *TYB* that are required for transposition. *TYA* encodes components of a virus-like particle that facilitates transposition while *TYB* is responsible for reverse-transcriptase, protease, integrase and RNaseH functions required in transposition. *TYB* is synthesized from a *TYA–TYB* transcript by translational frameshifting. Ty1 generates a 5 bp target-site duplication. (c) **Retroposons: I factor of *Drosophila*.** Retroposons are unusual in that they do not have terminal repeats. The left end of the I factor consists of the sequence CAG, and the right end is marked by the repeated sequence (TAA)$_4$. The element contains two open reading-frames, ORF1 and ORF2. Sequence analysis suggests that ORF2 encodes the reverse-transcriptase of the element. A 12 bp target-site duplication is made as a consequence of transposition. (d) **Fold Back elements: FB of *Drosophila*.** FB elements are diverse and the figure is drawn to illustrate their general structure. The ends of FB elements consist of a 10 bp repeat, (A)$_x$, followed by a 20 bp repeat that includes the 10 bp sequence, (AB)$_y$, followed by a 31 bp repeat that includes the 20 bp sequence, (ABC)$_z$. The figure shows only three of each repeat, but real elements contain variable numbers of each class of repeat and the numbers of repeats at both ends varies, so the left- and right-terminal inverted repeats are of different lengths. The 31 bp repeat is the most common and is repeated several tens of times. Not all repeats are perfect copies and some non-repeated spacer DNA of variable length occurs within the 10 and 20 bp repeat regions. A 4 kb region of unique DNA that contains three open-reading frames, believed to be required for transposition, is often found between the repeats. Transposition is associated with a target-site duplication of 9 bp.

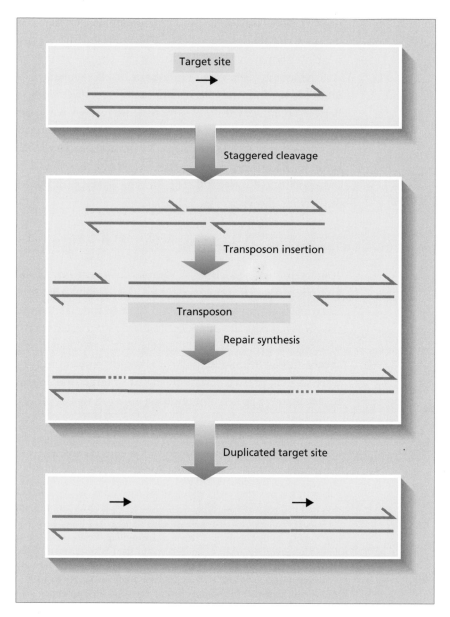

Fig. 5.3 Cleavage of the target-site to generate a duplication. A target-site in the top molecule is cleaved with a staggered break and a transposon is inserted by a pair of strand-transfers to the overhanging ends of the target. Repair DNA synthesis then copies the target at both sites of the inserted element to generate a duplication of the sequence originally present in the top chromosome. Note that the target-site duplication is generated by transposition but does not constitute part of the transposable element.

approximately to breaks on the same or opposite sides of the helix respectively.

By analysing a large number of insertion sites, it has been possible to determine whether certain elements have any preferences for target sequences. Most elements do have some preference for particular target sequences. In general, however, these sequences are short and degenerate, leading to an approximately random spectrum of insertions. An exception to this rule is the Tc1/Mariner subclass whose elements always insert at the sequence TA generating a 2 bp TA duplication.

5.5   Conservative and replicative transposition

Michel Faelen and Arianne Toussaint (1971, 1973) observed that transposition could be replicative when they showed that Mu mediated replicon fusion. As shown in Fig. 5.4, they obtained evidence that **co-integration** of λp*gal* and the bacterial chromosome could be mediated by bacteriophage Mu and that the fused replicons were separated by two directly repeated copies of the transposable element. The product of this replicon fusion is denoted a **co-integrate**. Fred Heffron, Dave Sherratt and their colleagues showed that mutant derivatives of the transposon Tn3 could also form co-integrates. These mutants lacked either a functional **resolvase** gene or the site (*res*) upon which resolvase acts and it was possible to conclude that Tn3 transposition was normally replicative but that a site-specific recombination system exists to resolve co-integrates to the final transposition products (Fig. 5.5). By contrast to the replicative transposition of Mu and Tn3, the transposon Tn10, which is a composite class I element, was shown by Nancy Kleckner and her colleagues to transpose via a non-replicative or **conservative** reaction (Fig. 5.6). *In vitro* evidence has also suggested that Mu can transpose by either replicative or conservative mechanisms and this gives rise to the pathway summary for the three elements, Mu, Tn3 and Tn10, shown in Fig. 5.7.

The transposition of the eukaryotic retrotransposons and retroposons is necessarily replicative since it is a copy of the element that is inserted at the new site. This is because transposition proceeds via an RNA intermediate as described in Section 5.8. However, this type of replicative transposition is different from bacterial replicative transposition in that the donor DNA is never attached to the target. Eukaryotic transposons such as Ac of maize and P of *Drosophila* may transpose by a variety of replicative and conservative mechanisms, as do the bacterial elements. Ac is believed to use a conservative mechanism because transposition can often be correlated with loss of the element from the donor site. This is shown in Fig. 5.8. Loss of P from the donor site (precise excision) occurs at a much lower

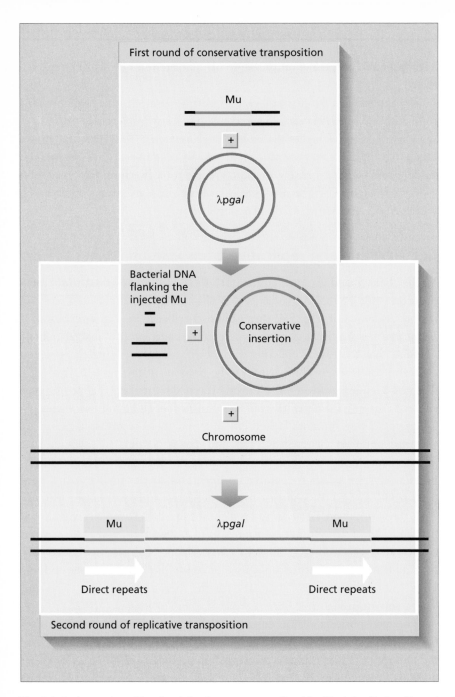

Fig. 5.4 Co-integration of lpgal and the chromosome mediated by Mu (after Faelen, Toussaint & Couturier, 1971, and Faelen & Toussaint, 1973). λpgal is a plaque-forming derivative of bacteriophage λ carrying the *gal* genes of *E. coli*. Gal⁻ cells were infected with this phage under conditions where the λpgal DNA was injected and re-circularized but was unable to replicate, to undergo homologous recombination or to integrate into the host chromosome. It was no surprise to discover that under these conditions no Gal⁺ transductants were formed. However, Gal⁺ transductants were obtained if the cells were co-infected with λpgal and Mu. The Mu must have mediated the formation of Gal⁺ transductants. When these Gal⁺ transductants were examined, it was shown that the λpgal DNA was sandwiched between two directly repeated copies of the Mu prophage. At the time the experiments were performed, it was not clear how this unusual structure was formed. However, now we understand this to be the result of two rounds of transposition. The first is a conservative integration of Mu into λpgal or the chromosome, followed by a second replicative round, fusing the two circles to form a co-integrate.

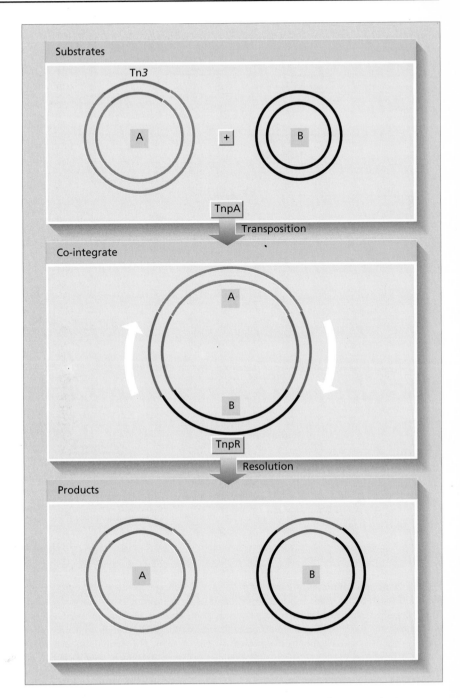

Fig. 5.5 Replicative transposition of Tn3. Tn3 transposes by a replicative mechanism generating a co-integrate. A site-specific recombination system then resolves the co-integrate into the products: the original replicon and the target replicon with a copy of the transposon. The first step is catalysed by the transposase encoded by the *tnpA* gene and the second by the resolvase encoded by *tnpR*.

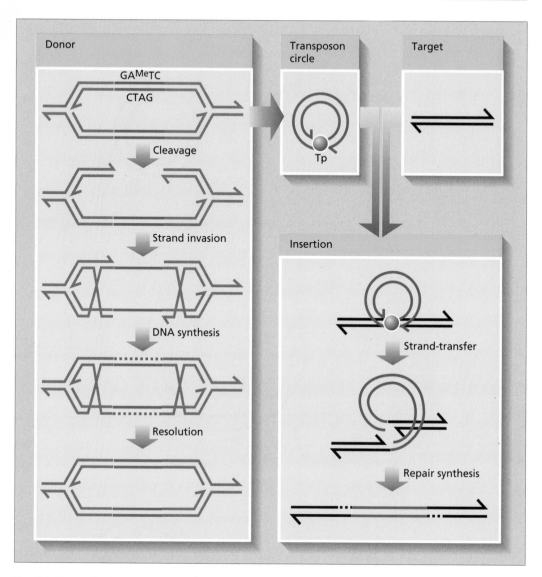

Fig. 5.6 Non-replicative transposition of Tn10. Tn*10* transposes via a non-replicative (conservative) mechanism. This mechanism is also called 'cut and paste', because the transposon is cut out of the donor replicon and pasted into the recipient. In the first step, the ends of the element are brought together and double-strand breaks are made by the transposase at the junction of the transposon with flanking sequences. This generates a transposon-circle, held together by the transposase protein (Tp). The transposase then makes staggered breaks in the target DNA and strand-transfer is completed. At first sight, this might seem to be a wasteful, destructive mechanism of transposition. However, two factors reduce its potentially negative consequences. Firstly, if the donor molecule is a multi-copy plasmid, the loss of one copy is rapidly made up by the copy-control mechanism to restore the original copy number. Secondly, transposition of Tn*10* is more efficient when the DNA is hemi-methylated at GATC sequences (the target for the *E. coli* Dam methylase) and this situation only occurs just after the element has been replicated. This means that the double-strand breaks introduced upon transposition result only in the destruction of, at most, one arm of a replication fork and probably, in fact, less than this since homologous recombination can regenerate the arm from which the transposon has moved. These events are shown diagrammatically in the figure.

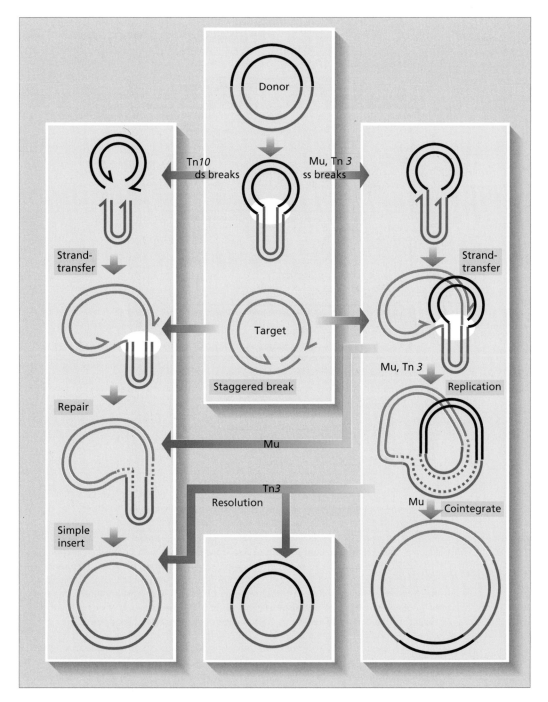

Fig. 5.7 Summary of prokaryotic transposition pathways. The transposition of Tn*10*, Tn*3* and Mu are compared to point out their similarities and differences. Initially, the transposase protein binds to the ends of the element and these are brought together into a complex ready for transposition. This **donor complex** is then cleaved (*Continued* on p. 122)

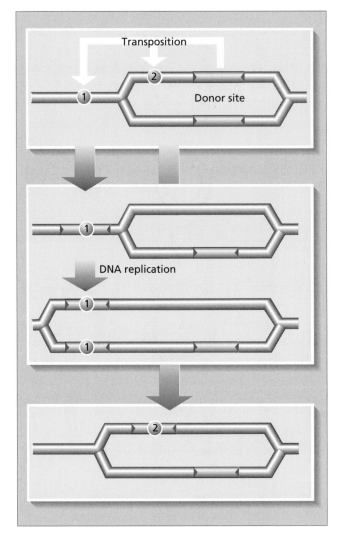

Fig. 5.8 Non-replicative transposition of Ac. Transposition is accompanied by loss of a copy of Ac on one sister chromatid at the donor site. If the target-site is in unreplicated DNA (1) transposition results in copies of Ac at the target-site on both sister chromatids. Alternatively, if the target-site is within replicated DNA (2), transposition results in a copy of Ac at the target-site on one sister chromatid. In both cases, the loss of one copy of the element at the donor-site suggests a conservative mechanism of transposition.

Fig. 5.7 (*Continued*) by transposase to generate either double-strand breaks (Tn*10*) or single-strand nicks (Tn*3* and Mu). Target DNA is cleaved with staggered breaks and strand-transfer proceeds. Following strand-transfer, DNA synthesis is initiated at the 3′ ends generated by cleavage of the target-site. This results in repair of the small gaps in the conservative pathway and in duplication of the element in the replicative pathway. Finally, in the case of Tn*3*, the co-integrate generated by replicative transposition is resolved into two products by site-specific recombination. The polarity of strand-transfer has been established for Mu and Tn*10*. The polarity for Tn*3* is assumed to be the same.

frequency than transposition, suggesting some form of replicative trans- position. However, the genome rearrangements mediated by P are not consistent with a replicative mechanism such as that used by Mu (see Section 5.9). A conservative mechanism tied to the pre-replication of the element and double-strand break repair (as is the case for Tn*10* transposi- tion) is a likely explanation of this apparent contradiction.

5.6 Strand-transfer in transposition

Strand-transfer is the formation of new phosphodiester-bonded connections and is a necessary consequence of transposition. In replicative transposition we can ask the question: What is the order of strand-transfer and DNA replication? This has led to the proposal of a plethora of models which can be subdivided into three groups: Firstly, models where two strand-transfers occur before replication; secondly, models where one strand-transfer occurs before replication; and thirdly, models where no strand-transfer occurs before replication. These three possibilities are shown diagrammatically in Fig. 5.9. In the case of Mu, Kiyoshi Mizuuchi and colleagues have

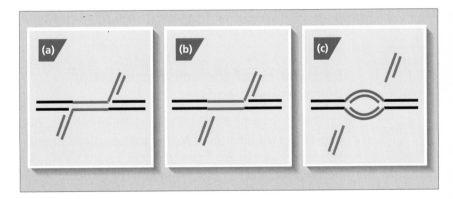

Fig. 5.9 Models for replicative transposition. (a) **Two strand-transfers prior to DNA synthesis.** Breaks are made at the ends of the transposable element and the cleaved ends are ligated to cleaved target-site. This generates two replication forks that can be used to replicate the element. This mechanism has been shown to be correct for the transposition of Mu and for retroviral integration. (b) **One strand-transfer prior to DNA synthesis.** A break is made at one end of the transposable element and this is ligated to one end of the cleaved target-site. DNA synthesis can then proceed unidirectionally through the element. After DNA synthesis, another round of cleavage and strand-transfer completes the replicative transposition reaction. A version of this mechanism has been proposed to account for retroposon transposition (see Fig. 5.15 for details). (c) **DNA synthesis prior to strand-transfer.** The transposable element is first replicated and then the ends are cleaved and ligated to the target. The transposition of Tn*10*, which is intrinsically non-replicative, can nevertheless be considered to proceed via a version of this type of replicative reaction by directing transposition to occur after host chromosome replication (as shown in Fig. 5.6).

shown that strand-transfer occurs between the 3′ end of the cleaved element and the 5′ end of the cleaved target-site to give an intermediate where two strand-transfers have occurred before replication. The nature of this intermediate was originally proposed by Jim Shapiro and is often referred to as the **Shapiro intermediate**. These experiments are described in Fig. 5.10.

Bob Craigie, Kiyoshi Mizuuchi, Ron Plasterk and their colleagues have gone on to show that Mu transposition and HIV integration proceed via two chemical reactions. The first is cleavage of the phosphoester bond joining the element to its flanking DNA to generate a 3′OH at the end of the element. The second is transesterification using the 3′OH at the cleaved end of the element as a nucleophile to attack a phosphoester bond at the target. These reactions are shown diagrammatically in Fig. 5.11. There is no covalent protein-DNA intermediate in either reaction, in contrast to the situation seen in site-specific recombination (see Chapter 4).

All transposable elements that have been studied so far, transpose via a similar mechanism. The elements are cleaved precisely at the 3′ end to leave 3′OH residues. These 3′OH residues are used to attack the target-site and complete the first round of strand-transfers. The 5′ ends of the elements are not always cleaved and if they are, this is not always at the same position. A summary of the 3′ and 5′ cleavage sites for six elements is shown in Fig. 5.12. The precise location of the 3′ cleavage sites ensures that the integrity of the elements is preserved.

5.7   Recognition of the active orientation of transposon ends

The ends of transposable elements are required to be oriented correctly with respect to each other in order for transposition to occur. Furthermore, for bacteriophage Mu we know that the correct orientation of ends is also required for any strand-transfer to be detected *in vitro*. Craigie and Mizuuchi therefore asked whether it is the orientation of the ends *per se* on a DNA molecule or their orientation in supercoiled DNA that is the feature responsible for their correct recognition. Their experiment, shown in Fig. 5.13, demonstrated that it is the relative orientation of the ends in the supercoiled structure that is required for the activation of strand-transfer.

5.8   Transposition via an RNA intermediate

The eukaryotic retrotransposons and retroposons transpose via an RNA intermediate. Direct evidence for an RNA intermediate in the transposition of the yeast retrotransposon, Ty, was obtained in Gerald Fink's laboratory

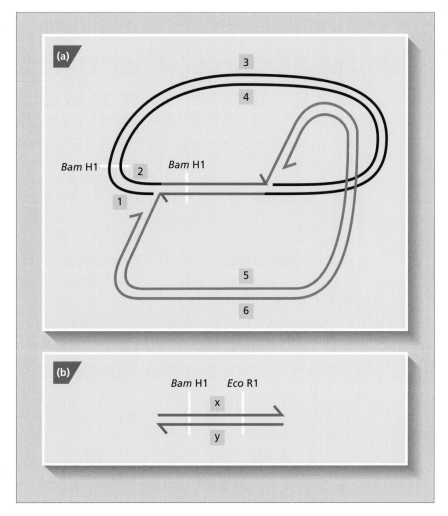

Fig. 5.10 Strand-transfer in bacteriophage Mu transposition. After Mizuuchi, 1983 &
1984, and Craigie & Mizuuchi, 1985. (a) **Structure of the strand-transfer product.**
Transposition was carried out *in vitro* between a Mu-containing plasmid (red and black)
and a circular target molecule containing no Mu sequences (blue). This reaction was
performed under conditions that did not permit any DNA synthesis, in order to isolate the
intermediate formed by strand-transfer in the absence of replication. This intermediate is
shown in the figure and its structure was deduced by observing the presence of all the
single-strand products predicted after digestion with the restriction enzyme *Bam*HI. These
fragments are labelled 1–6 and it can be seen that fragments 5 and 6 are characteristic of
strand-transfer at both ends of the element. The arrowheads at the ends of the element
show the position of the 3′ ends which are ligated to the 5′ ends of the cleaved target. (b)
Polarity of strand-transfer. The single-strand fragments x and y define the two strands of
Mu and were used to determine the strandedness of fragments 3 and 6, in (a), by
hybridization. Because x hybridized to 3 and y hybridized to 6, it was clear that the 3′ end
of the transposon was joined to the 5′ end of the cleaved target-site.

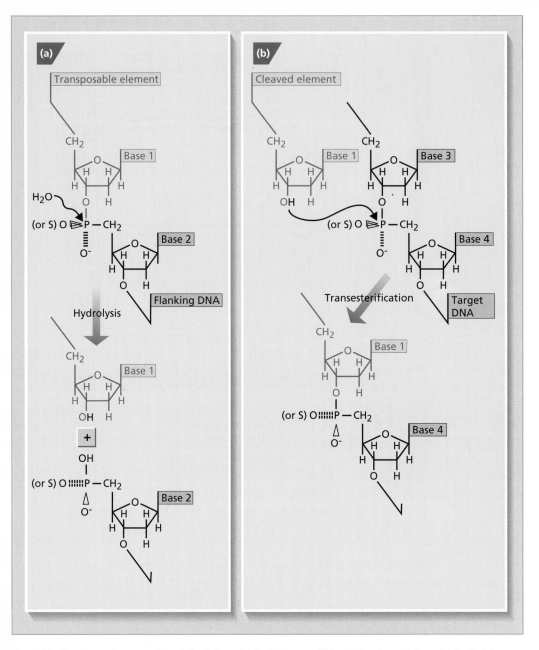

Fig. 5.11 Chemistry of transposition (after Mizuuchi & Adzuma, 1991 and Engelmen, Mizuuchi & Craigie, 1991. Transposition occurs in two separate reactions shown in parts a and b of the figure). (a) Firstly, hydrolysis of the phosphoester bond adjacent to the last nucleotide of the transposable element (shown in red) generates a strand-break with a free 3′OH at the end of the element. This reaction is believed to occur by nucleophylic attack of the phosphoester bond joining the transposon to its flanking DNA. H_2O acts as the nucleophile and the transposase makes the phosphoester bond more susceptible to nucleophylic attack. (b) Secondly, the 3′OH that has been generated at the end of the element is joined to the target-site to complete strand-transfer. In this case, the nucleophile is the 3′OH of the cleaved transposon and the attack of the phosphoester bond results in transesterification. Again, it is thought that the transposase makes the phosphoester bond more susceptible to nucleophylic attack. Both the hydrolysis and the transesterification reactions (a and b) are accompanied by inversion of the chirality of the phosphate as demonstrated in experiments where sulphur atoms, shown as (S), were substituted for oxygen to enable the chirality to be monitored. Inversion of chirality is characteristic of a

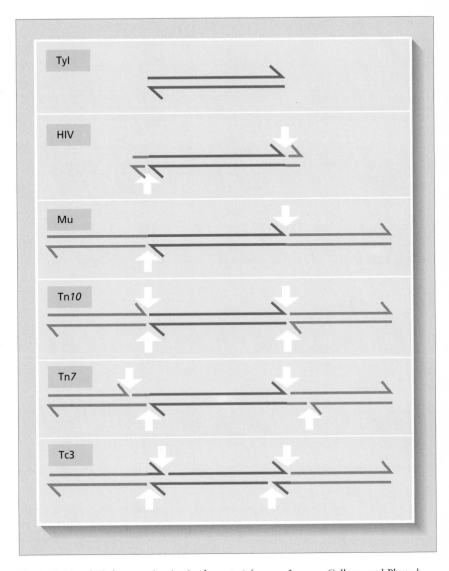

Fig. 5.12 3′ and 5′ cleavage sites in six elements (after van Luenen, Colloms and Plasterk, 1994). For all six elements, the 3′ cleavage sites are precisely at the end of the element. The 5′ cleavage sites, on the other hand, differ. Ty1 is reverse transcribed to make a product that is precisely the length of the element. HIV has 2 bp of flanking DNA that is cleaved only on the 3′ sides. Mu is also cleaved only on the 3′ ends. Tn10 is cleaved on both strands precisely at the ends. The 5′ cleavage sites of Tn7 lie three bases from the end, within the flanking DNA; while those of Tc3 lie two bases from the end, within the element itself. In all cases, the 3′ end of the element is joined to the 5′ end of the cleaved target ensuring that the element is transposed precisely and that a target-site duplication is generated.

Fig. 5.11 (*Continued*) reaction that occurs in one (or any odd number) of steps. Therefore, it is believed that transposition does not involve the formation of a protein DNA intermediate in strand-transfer as is the case for the site-specific recombination reactions described in Chapter 4.

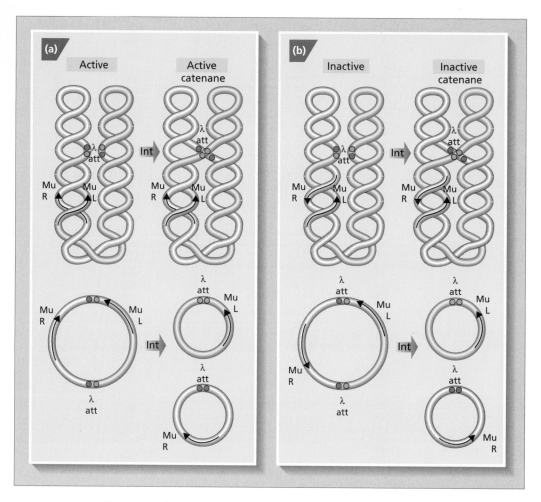

Fig. 5.13 End recognition in bacteriophage Mu transposition. (After Craigie and Mizuuchi, 1986.) The proficiency of the Mu ends for strand-transfer was compared as a function of their relative orientation within a supercoiled structure. The connectivity of the DNA double-strands was then altered by Int-mediated site-specific recombination (see Chapter 4). Catenated molecules, where the two ends are correctly oriented with respect to each other in the supercoil but are located on the two different partners of the catenane, are shown in (a). Catenated molecules with incorrectly oriented ends are shown in (b). A knotted molecule with correctly oriented ends is shown in (c). Finally, a knotted molecule with incorrectly oriented ends is shown in (d). In all cases, the proficiency of the Mu ends to carry out strand-transfer was found to be determined by the relative orientation in the supercoiled structure and did not depend on the connectivity of the DNA strands.

using a clever strategy. An intron was inserted into a Ty derivative and was shown to be spliced out in the majority of transposition products. Since splicing only occurs in RNA, an RNA intermediate must have existed. This experiment is described in Fig. 5.16.

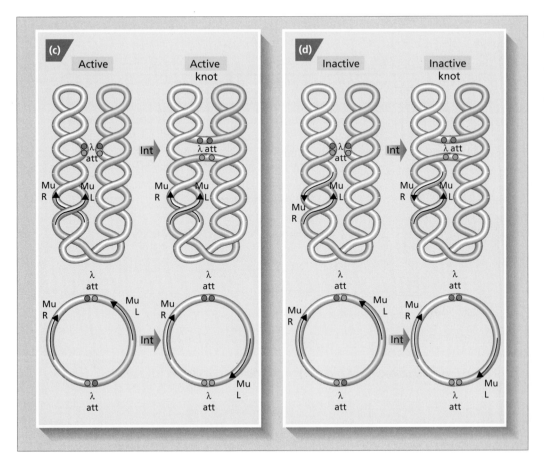

Fig. 5.13 (*Continued*)

The retrotransposons use a mechanism (shown in Fig. 5.14) analogous to that of retrovirus replication. The integron reaction proceeds by a mechanism similar to Mu strand-transfer, by nucleophylic attack of the target-site by the 3'OH ends present at the end of the cleaved element.

Less is known about retroposon transposition. The structure of the elements and the homologies encoded to reverse transcriptase strongly suggest the existence of an RNA intermediate. However, the absence of long terminal repeats (LTRs) requires a different mechanism of transposition to that employed by retrotransposons. Indeed they are believed to transpose via an RNA intermediate that is directly copied into DNA during integration (as shown in Fig. 5.15).

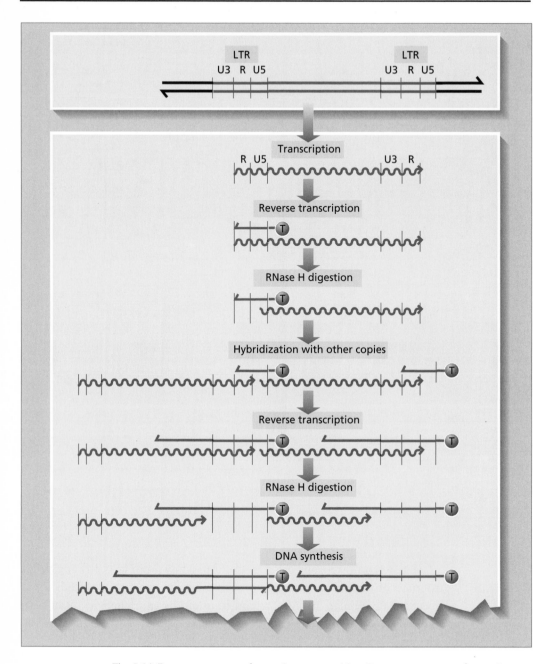

Fig. 5.14 Retrotransposon and retrovirus transposition. Retrotransposons and retroviruses contain directly repeated long terminal repeats (LTRs) that can be subdivided into regions U3, R and U5. Transcription generates an RNA molecule (wavy line) containing directly repeated copies of R. A host tRNA (designated T) then binds to the transcript, just to the right of U5. This is used as a primer for reverse-transcription that proceeds through R. RNase H digestion then removes part of the RNA that has just been copied. The overhanging DNA strand can then be used to bridge between copies of the transcript by hybridization to the R region that was originally present at both ends of the RNA. This

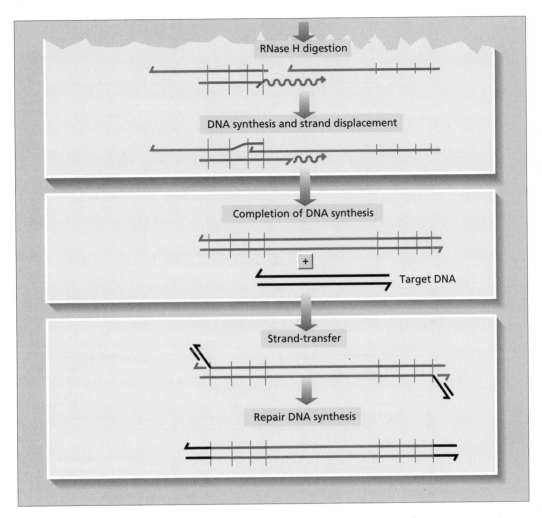

Fig. 5.14 (*Continued*) allows reverse-transcription to extend the DNA copy through the element. A combination of RNase digestion and DNA synthesis ensures the removal of the RNA and its replacement with DNA. When DNA synthesis has been completed, there is a complete double-strand copy of the element. In some cases, this includes several bp flanking it on either side (see Fig. 5.12). Strand-transfer of the 3′ end of the element to the 5′ overhang of a cleaved target-site then generates an intermediate similar to that generated in Mu transposition. This proceeds to the product by repair DNA synthesis. Retroviruses differ from retrotransposons in that copies of the RNA genome are packaged into infective viral particles.

5.9   Transposition can mediate genetic rearrangement

Transposons are a major cause of genetic rearrangement. They can mediate deletions, inversions and fusions imparting considerable flexibility to the organization of the genome over an evolutionary time scale. Like mutation, genome rearrangement has both a cost and a benefit. Most rearrangements

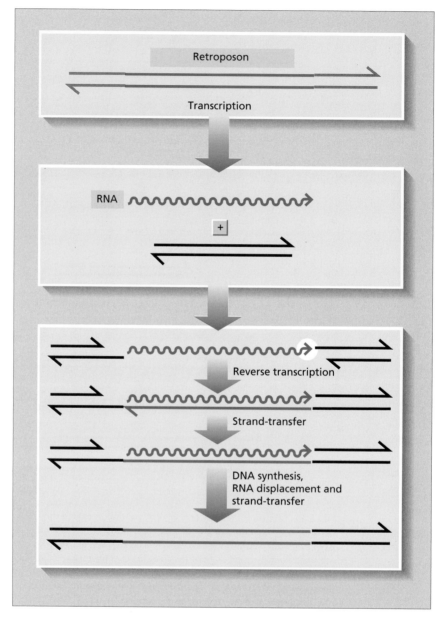

Fig. 5.15 Retroposon transposition. Retroposon transposition is believed to proceed by a mechanism such as that shown. The element is transcribed into RNA. The 3′ end of this RNA then interacts with a cleaved target-site. This is followed by reverse-transcription of the element to generate a DNA–RNA hybrid. The 3′ end of the hybrid is then joined by strand-transfer to the target. DNA synthesis and RNA strand-displacement completes the transposition of the element.

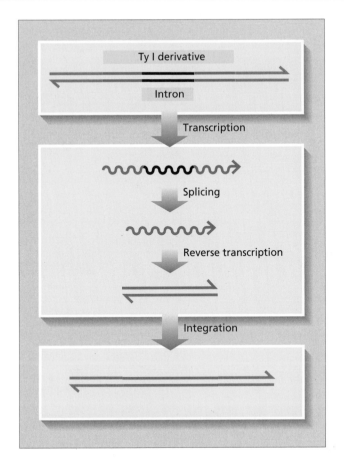

Fig. 5.16 Evidence for transposition via an RNA intermediate. (After Boeke, Garfinkel, Styles and Fink, 1985). Direct evidence for the transposition of Ty via an RNA intermediate was obtained by an intron-loss experiment. DNA containing an intron was inserted into a Ty derivative and the products of transposition analysed to determine whether or not they had lost the intron. Loss of the intron would indicate the presence of an RNA intermediate, since splicing only occurs on an RNA substrate. The intron had indeed been lost in most of the new insertions, confirming directly that an RNA intermediate must have been involved in transposition.

are likely to be detrimental to the individual in which they have occurred but the potential to reorganize the genome is essential for chromosome evolution. Figure 5.17 summarizes some of the genetic rearrangements that can occur as a consequence of transposition. Since the rearrangements are caused either as a consequence of normal or abortive transposition, not all elements mediate all types of event. For example: **precise excision** can be detected for Tn*10* but not for Tn*3* or Mu. Ac and Ds often excise imprecisely to leave two copies of the target-site duplication at the original site of insertion. Tn*10* and P can excise imprecisely by strand-slippage during replication between short direct repeats to leave parts of the element behind.

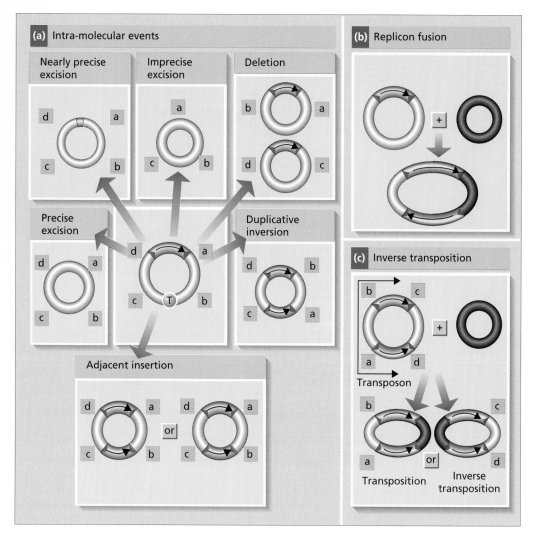

Fig. 5.17 Genetic rearrangements mediated by transposition. A selection of intra-molecular rearrangements mediated by transposable elements are shown in (a). The transposable element is shown in an arbitrary orientation relative to four sections of a circular chromosome marked **a**, **b**, **c** and **d** and a target-site is marked **T** so that the consequences of rearrangement are clear. Genome fusion (or co-integration) is shown in (b) and the inter-molecular transposition of a compound transposon is shown in (c). In this last situation we may consider the transposon to be the section of DNA containing markers **a** and **b** or that containing **c** and **d**. The transposition of the section not considered to be 'the transposon' is sometimes called **inverse transposition**.

Mu mediates **deletions** and **duplicative inversions** as a consequence of replicative transposition whereas Tn*10* mediates adjacent insertions when a copy of the element transposes from one copy of the chromosome to another. Only composite transposons mediate **inverse transposition.**

As an example of how the mechanism of transposition can lead directly to rearrangement, the formation of duplicative inversions and deletions by replicative transposition is shown in Fig. 5.18. These classes of rearrangement are often observed for prokaryotic elements, but many of the rearrangements generated by eukaryotic transposons, such as Ac, Ds or P, are not the direct consequences of transposition. Instead, they are due to secondary processing of intermediates or products of transposition. For example, certain Ds elements are frequent sites of chromosome breakage. The broken ends can then fuse by illegitimate end-joining to generate abnormal structures such as dicentric chromosomes that can enter breakage-fusion-bridge (BFB) cycles (see Chapter 6) resulting in further rearrangements.

5.10 ## Conclusion

Despite their diversity, transposable elements share some remarkable features. The most notable of these is the similarity of the underlying chemistry of strand-transfer. In all cases that have been determined, transposition proceeds in two steps. A hydrolysis reaction generates a 3′OH at the end of the element and this attacks the phosphodiester backbone of the target directly to generate the strand-transfer product. In almost all cases this generates a target-site duplication at the site of insertion. It is likely that this is because of the double-helical nature of DNA, coupled to the action of proteins that recognize particular faces or grooves of the helix, thus displacing cleavage points with respect to each other. Most elements share the structural feature of inverted terminal repeats. This can be explained by the requirement of a protein that recognizes the ends to initiate strand-transfer. In order for this to occur precisely, the transposase must firstly recognize the ends of the element which implies the existence of protein-binding sites at both ends, and secondly strand-transfer must occur precisely at the last base pair of the element, respecting the polarity of DNA. This is facilitated if the last base of the strand that is cleaved at one end is the same as that adjacent to the cleavage point on the other strand at the other end of the element. Both these requirements are simply fulfilled by the existence of inverted repeats.

Argument flourishes as to whether transposable elements confer a selective advantage to their host or whether they can be considered as molecular parasites (pieces of selfish DNA acted upon directly by evolution). Some elements clearly do confer a selective advantage to their host under particular circumstances (e.g. drug resistance in the presence of an antibiotic) but others would appear only to confer a genetic load. However, chromosomal rearrangement is essential for evolution and transposable

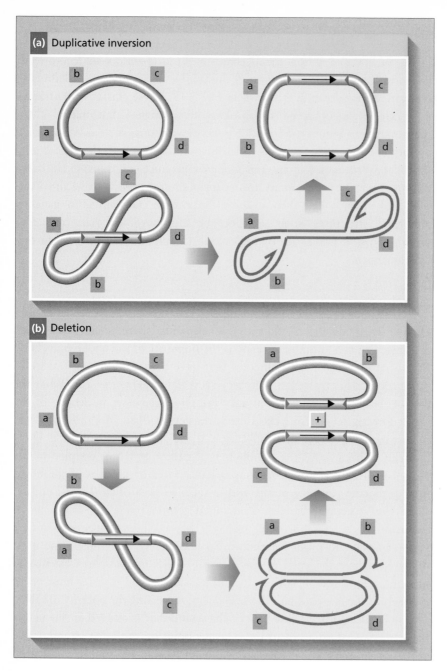

Fig. 5.18 Formation of duplicative inversions and deletions by replicative transposition. (a) **Duplicative inversions.** One simple consequence of intra-molecular replicative transposition is duplicative inversion. Insertion is made at a target-site between b and c to give rise to the 'Shapiro intermediate' shown. This is resolved by replication into the duplicative inversion. The two copies of the transposable element are in inverted repeat with respect to each other and the backbone sections ab and cd are inverted with respect to their relative orientation in the substrate. (b) **Deletion.** Insertion is again into a site between b and c but in the opposite orientation to that shown in (a). The consequence of this orientation of insertion is the alternative form of the 'Shapiro intermediate'. This can then be resolved by replication into two circular molecules, each containing one copy of the transposable element. If one of these circles contains an origin of DNA replication and the other does not, the former will be inherited and latter lost.

elements do provide mechanisms for such change. Perhaps we should think of the relationship as symbiotic rather than parasitic.

Further reading

Berg D. E. & Howe M. M. (1989). Thirty review articles on transposable elements in *Mobile DNA*. American Society for Microbiology, Washington.

Mizuuchi K. (1992). Transpositional recombination: mechanistic insights from studies of Mu and other elements. *Ann. Rev. Biochem.* **61**, 1011–1051.

References

Boeke J. D., Garfinkel D. J., Styles C. A. & Fink G. R. (1985). Ty elements transpose through an RNA intermediate. *Cell* **40**, 491–500.

Craigie R. & Mizuuchi K. (1985). Mechanism of transposition of bacteriophage Mu: Structure of a transposition intermediate. *Cell* **41**, 867–876.

Craigie R. & Mizuuchi K. (1986). Role of topology in Mu transposition. Mechanism of sensing the relative orientation of two DNA segments. *Cell* **45**, 793–800.

Engelman A., Mizuuchi K. & Craigie R. (1991). HIV-1 DNA integration: mechanism of viral DNA cleavage and DNA strand transfer. *Cell* **67**, 1211–1221.

Faelen M. & Toussaint A. (1973). Connecting two unrelated DNA sequences with a Mu dimer. *Nature New Biol.* **242**, 1–4.

Faelen M., Toussaint A. & Couturier M. (1971). Mu-1 promoted integration of a λ-gal phage in the chromosome of *E. coli*. *Molec. Gen. Genet.* **113**, 367–370.

Mizuuchi K. & Adzuma K. (1991). Inversion of the phosphate chivality at the target site of the Mu DNA Strand transfer: evidence for a one-step transesterification mechanism. *Cell* **66**, 129–140.

Mizuuchi K. (1983). *In vitro* transposition of bacteriophage Mu: a biochemical approach to a novel replication reaction. *Cell* **35**, 785–794.

Mizuuchi K. (1984). Mechanism of transposition of bacteriophage Mu: polarity of the strand transfer reaction at the initiation of transposition. *Cell* **39**, 395–404.

Sherratt D. (ed.) (1995) *Mobile Genetic Elements: Frontiers in Molecular Biology*. Oxford University Press, Oxford.

van Luenen H. G. A. M., Colloms S. D. & Plasterk R. H. A. (1994). The Mechanism of Transposition of Tc3 in *Caenorhabditis elegans*. *Cell* **79**, 293–301.

6 Illegitimate Recombination

6.1 Introduction

Illegitimate recombination events occur at DNA sequences that share little or no homology with each other. These reactions are, however, more promiscuous than site-specific recombination reactions, occurring at many different sites and DNA sequences. They are likely to be the most primitive form of recombination, not requiring the evolution of complex systems for the recognition of specialized sequences or the evolution of a mechanism for recognizing DNA homology. These reactions can be extremely efficient. They are also central to the mechanisms of carcinogenesis, inherited disease and genome evolution.

Illegitimate recombination can be subdivided into two primary mechanisms. The first of these is **end-joining**. This is a reaction that results in the ligation of broken DNA ends to each other. The second is **strand-slippage**, where DNA replication skips from one template to another, resulting in recombination. In addition to these primary mechanisms, there are events that occur as a result of aberrant recombination, replication and topoisomerase reactions. Illegitimate recombination can result in many different products such as, frameshifts, deletions, inversions, fusions and DNA amplifications.

The two primary mechanisms of illegitimate recombination can occur at a wide variety of DNA sequences. This promiscuity makes them both a threat to the integrity of genomes and a powerful mechanism for their evolution. They also lead to problems for homologous targeting in mammalian cells and the stable cloning of eukaryotic DNA in bacteria.

The reactions that lead to immunoglobulin (Ig) gene rearrangements have been included in this chapter since they have features that resemble illegitimate recombination. They are, however, significantly different from other illegitimate events in that they are focused at particular sites. They could have been included in the chapter on site-specific recombination or (in the case of class-switching) in the chapter on homologous recombination. Until more is known on the mechanism of these events their classification is difficult.

6.2 End-joining

End-joining is a very efficient reaction in eukaryotic cells. This was first recognized by Barbara McClintock in the early 1940s. She observed that if a maize cell inherits a broken chromatid and this chromatid is allowed to go through replication, the two broken ends will fuse very efficiently to give rise to a dicentric chromatid. This chromatid will then break at mitosis and regenerate broken ends that will fuse again to regenerate a new dicentric chromatid. This will then break at the next mitosis and the cycle of **breakage–fusion–bridge** (or **BFB**) will continue for many rounds until telomeres are added to the broken ends (Fig. 6.1). BFB cycles have been shown to be responsible for the amplification of certain chromosomal regions in a head-to-head manner, generating megabase (mb) palindromes that are nearly perfect at their centres. This mechanism of palindrome formation is shown in Fig. 6.1.

The mechanism of end-joining has been extensively studied by David Roth and John Wilson using a model system in monkey cells and Petra Pfeiffer using a similar system in *Xenopus* egg extracts. As shown in Fig. 6.2, end-joining can occur by directly abutting the broken ends and ligating the strands. However, if the strands cannot be ligated (for instance if they have the same polarities of end) they can be paired and then the ends filled-in. End-joining results in the formation of new joints, irrespective of the DNA sequence of the ends, and does not require any homology between the recombining partners. If there has been pairing at micro-homologies, the reaction will appear to have occurred at short direct repeats, typically of 2–5 base pairs (bp). Furthermore, in about 10% of cases random nucleotides are found located between the broken ends.

End-joining reactions in *Escherichia coli* require short stretches of DNA homology for pairing to occur. This is because *E. coli* is not able to ligate single-strand ends. When linearized plasmid molecules are introduced into cells they can recircularize using the ends generated by linearization. If this reaction does not occur, then two other reactions are possible. One of the ends can recombine with an internal sequence that has a small run of homologous bases, or two internal sequences that share a small region of homology can be used. It is believed that the regions of short homology may be found after nuclease digestion to reveal single-stranded DNA.

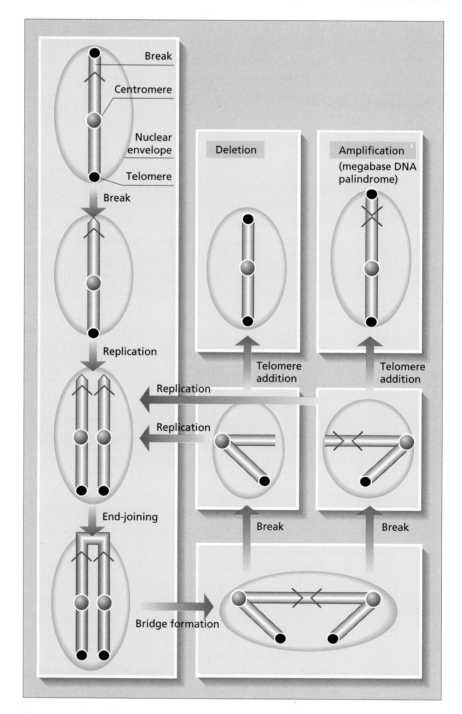

Fig. 6.1 Breakage–fusion–bridge (BFB) cycles can generate amplified DNA sequences arranged as long DNA palindromes. A chromosome is shown which undergoes breakage between a genetic marker (denoted by an arrowhead) and the telomere. At mitosis, the chromosome is replicated to generate two broken chromatids. These fuse efficiently by an

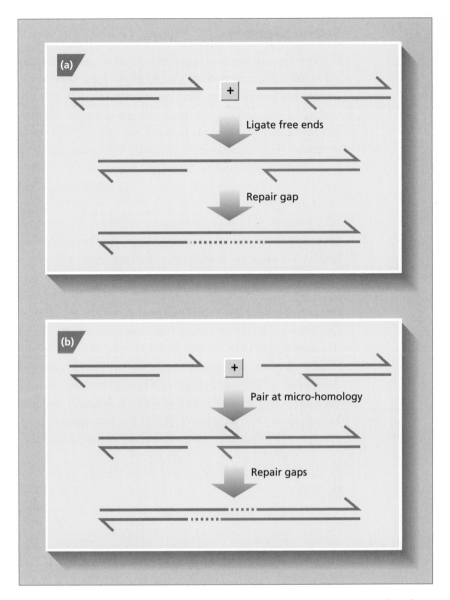

Fig. 6.2 End-joining reactions. End-joining in mammalian cells can proceed by either of two pathways. The two ends can either be ligated to each other as shown in (a) or can pair by hydrogen-bonding prior to ligation as shown in (b). In both cases, a small amount of DNA synthesis is required to complete both strands.

Fig. 6.1 (*Continued*) end-joining reaction. At anaphase of mitosis, the dicentric chromatid is stretched between the two poles of the dividing nucleus. This bridge breaks and the two products segregate to the daughter nuclei. Breakage is likely to be asymmetric and generates one deleted and one amplified chromosome. The amplified chromosome contains a long palindrome that was formed at the time of end-joining. Both the deleted and amplified chromosomes remain broken and can re-enter the BFB cycle. Cycles continue until a telomere is added to a broken end.

Strand-slippage

In 1966 George Streisinger proposed that frameshift mutations could occur by a mechanism involving the mispairing of a newly synthesized DNA strand and its template. This same mechanism, when applied on a larger scale, can account for deletions and duplications of DNA whose length is determined by the distance between the old and the new templates. This is shown in Fig. 6.3. Because this mechanism involves the pairing of a newly replicated strand with a template strand, these events are likely to be favoured by Watson–Crick base-pairing and therefore frequently occur at small direct repeats.

Certain directly repeated sequences seem to be hot-spots for strand-slippage. Studies in Jeffrey Miller and Barry Glickman's laboratories have identified a hot-spot for mutation in the *lacI* gene of *E. coli* that accounts for about two thirds of all spontaneous mutations. It consists of a DNA sequence 5′CTGG3′, or its complement 5′CCAG3′, that is repeated three times in tandem. Both increases and decreases in the number of copies of this tetra-nucleotide sequence have been observed, suggesting that looping out may occur both on the template and the newly polymerized strand. Similar deletion and amplification mutations in human chromosomes are associated with genetic disease. These mutations occur in tri- and tetra-nucleotide repeats and have been called **dynamic mutations.** In the fragile X syndrome, the sequence $(5′CCG3′)_n$ is amplified from 6–52 copies in normal individuals to more than 200 copies in affected individuals. In

Fig. 6.3 Deletion by strand-slippage. During DNA replication, the newly synthesized DNA may occasionally dissassociate from its template and hybridize to a complementary sequence downstream. Complementary sequences are shown as a,a′; b,b′; and c,c′. This results in deletion of DNA sequences.

Huntington's disease (HD), the sequence (5′CAG3′)$_n$ undergoes a more modest amplification from 6 to 33 copies to 35–121 copies. It is thought that the sequence (5′CXG3′)$_n$ may have the potential to form an unusual secondary structure when single-stranded that stabilizes the looped out strand. This may be a pseudo-hairpin, stabilized by G/C base-pairing, as shown in Fig. 6.4. In addition, it has been shown that these sequences are hard to replicate by DNA polymerases and this may contribute to a higher than normal pausing that also facilitates strand-slippage. A role for strand-slippage in dynamic mutation has not yet been proved and alternative models, in which increases and decreases in repeat numbers are due to homologous recombination, are also possible.

Both end-joining and strand-slippage reactions can result in the forma-

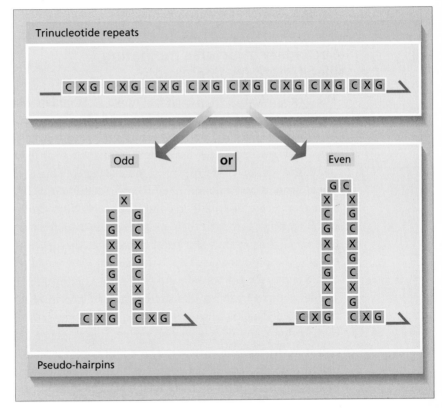

Fig. 6.4 Hypothetical pseudo-hairpins in a tri-nucleotide repeat. All unstable tri-nucleotide repeats associated with heeman inherited disorders share the sequence (5′ CXG 3′)$_n$. This could be explained if this sequence is able to form one or more unusual secondary structures stabilized by intra-strand base-pairing between C and G residues. The figure is drawn to emphasize the potential role of G/C bp in stabilizing structures composed of odd or even numbers of repeats. Evidence consistent with the formation of such pseudo-hairpin structures has recently been obtained in *in vitro* physical studies of oligonucleotides, and data consistent with a preference for folding into an even-membered structure has been obtained *in vivo* (see Darlow and Leach, 1995, and references therein).

tion of recombinant junction points at the positions of short direct repeats. Similarly, both mechanisms have the potential to form junctions at completely non-homologous sites, if we admit the possibility of non-Watson–Crick hydrogen-bonding interactions in the strand-slippage reaction. It is, therefore, not possible to distinguish these two mechanisms via the structure of their products. End-joining reactions involve the interaction of ends that have been generated by double-strand cleavage, whereas strand-slippage involves the interaction of only one single-strand end with an unbroken strand accompanied by DNA replication. The introduction of double-strand breaks *in vitro* before DNA is re-introduced to cells has been a useful experimental technique for studying end-joining. However, no such simple strategy is available to study strand-slippage and therefore many of the arguments for its participation in large-scale rearrangements have been indirect, as described in the Section 6.4.

6.4   Secondary structures promoting illegitimate recombination

The best studied DNA sequences known to stimulate illegitimate recombination are inverted repeats and long palindromic sequences. Palindromic sequences are, in fact, inverted repeats with no spacer DNA between the repeats. These sequences have the potential to form hairpin structures in single-strands, or cruciforms in double-strands. This is shown in Fig. 6.5. These structures stimulate illegitimate recombination by strand-slippage. Palindromes are deleted by recombination between short, directly repeated DNA sequences close to their ends. There is a preference for one repeat to be located just within the palindrome and the other to be located just outside the palindrome. This is consistent with a model in which replication enters the palindrome and only stalls after progressing some distance. The nascent strand then dissociates and can hybridize to any complementary sequence that is nearby. Such a sequence is typically found downstream of the palindrome. This asymmetry permits us to define a donor and a target repeat in the deletion reaction. The donor repeat is that located within the palindrome and the target repeat is that located outside (Fig. 6.6). This asymmetry has been used to test whether strand-slippage occurs more frequently on the leading or the lagging strands of the replication fork. This was done by determining the effect of inversion of a section of DNA that includes a palindrome and a favoured target-site (Fig. 6.7). The data suggests that deletion can occur more frequently on the lagging strand.

Inverted repeat sequences, with unique DNA between the repeats, also undergo deletion, but at a much lower frequency than perfect palindromes. These illegitimate recombination reactions account for the excision of trans-

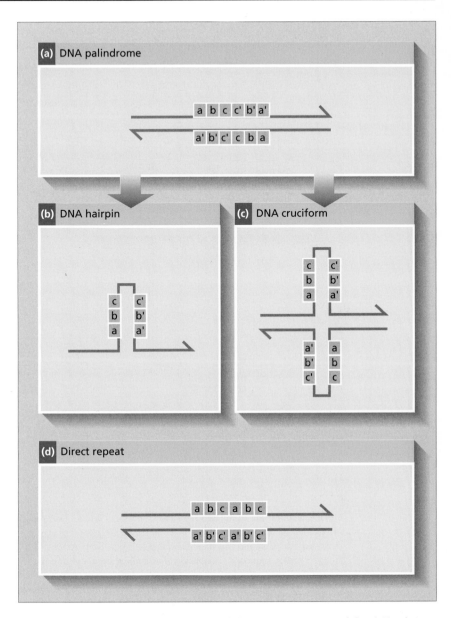

Fig. 6.5 Palindromes, hairpins, cruciforms and direct repeats. Because of the 5′–3′ polarity of DNA, only a double-strand can be palindromic (reads identically in one direction and in the other) and each individual strand is self-complementary. This self-complementarity means that intra-strand base-pairing is possible. In this figure, bases a, b and c are complementary to bases a′, b′ and c′ respectively. In (a) the palindrome is represented in its linear conformation. In (b) one strand of the palindrome has formed intra-strand bps to generate a hairpin structure. In (c) this has occurred on both strands to form a cruciform structure. A directly repeated sequence does not have the same potential for secondary structure formation, and is shown in (d).

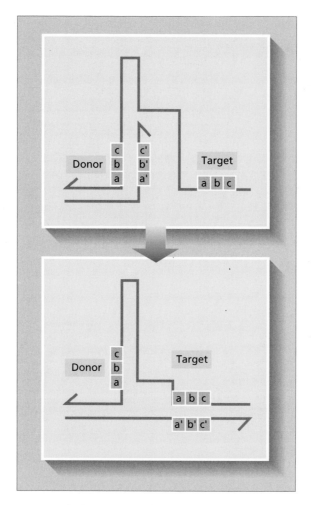

Fig. 6.6 Palindrome-stimulated replication-slippage. When DNA replication enters a region of intra-strand base-pairing, there is an increased likelihood of stalling. The 3′ end of the newly replicated strand can then dissociate and pair downstream with another complementary sequence. This means that stalling occurs within the DNA palindrome and the re-association to a new complimentary sequence occurs on the downstream side of the palindrome. This allows definition of a donor repeat (inside the palindrome) and a target repeat (outside the palindrome) that are involved in the strand-slippage reaction. These direct repeats (represented by a b c) can be as small as three nucleotides in length.

posable elements (e.g. Tn*10* and P) and probably proceed by a similar mechanism. Very long perfect palindromes (longer than approximately 150 bp in total length) cannot be replicated in *E. coli* at all. This is because of the action of SbcCD, an ATP-dependent exonuclease that can cleave DNA hairpin structures.

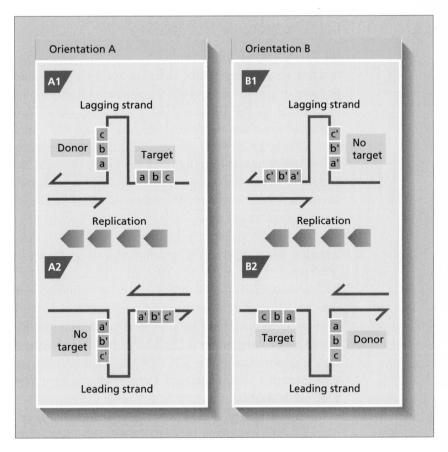

Fig. 6.7 Strand-slippage occurs preferentially on the lagging-strand (after Trinh & Sinden, 1991). As a newly replicated DNA strand enters a palindromic sequence in one direction, strand-slippage is favoured if a good target repeat is located downstream of the palindrome. A good target repeat is a sequence, just outside the palindrome, with several bases of complementarity to a sequence just within the palindrome. Since this preferred target repeat must be located on one side of the palindrome only, inversion moves it from leading to lagging strand or vice-versa. If a DNA palindrome and its surrounding sequence (including its preferred target for strand-slippage) is inverted, it is possible to determine whether deletion occurs preferentially on the leading or the lagging-strand. In the example shown, replication through the palindrome occurs from right to left. In orientation A, the preferred target is downstream of replication on the lagging strand (A1), whereas in orientation B the same preferred target is downstream of replication on the leading strand (B2). Orientation A should, therefore, show an elevated deletion frequency if strand-slippage occurs, preferentially on the lagging-strand. This predicted orientation-dependence is observed.

6.5

Immunoglobulin VDJ joining

Igs (or antibodies) are proteins that bind to a wide array of antigens and function to focus the body's defences against foreign material. In order to perform this function, Igs have a variable region that is responsible for the recognition of the antigens and a constant region responsible for the conserved functions required in the body's response to the foreign material (Fig. 6.8). In 1965, W. J. Dreyer and J. C. Bennett proposed that recombination might be involved in the generation of the wide diversity of variable regions observed, but it has not been until the last decade that clear evidence for this has been obtained. The mouse has been the most extensively used model system for the organization and rearrangement of the Ig genes, therefore this text concentrates on that system. Other well-studied systems, including those of chickens and humans, differ in detail but not in basic principles.

The organization of the three Ig loci of the mouse is shown in Fig. 6.9. One locus encodes the heavy chain and two encode the light chain. Each of these is composed of variable and constant parts which are brought together by recombination. The heavy chain locus rearranges first. Initially a D_H segment is joined to a J_H segment. Then one of the large number of V_H segments is joined to $D_H J_H$. These reactions are collectively known as VDJ recombination.

Each coding region (V, D or J) is adjacent to a signal sequence that is required for recombination. The signal sequences are initially brought together and recombination occurs at the junction between the signal and coding regions to generate a new coding junction and a signal junction (as shown in Fig. 6.10). The signal junction is normally relatively precise but can involve the addition of a few untemplated bases between the copies of

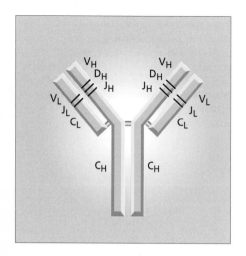

Fig. 6.8 Structure of the immunoglobulin (Ig) molecule. The Ig protein is composed of two copies of a light chain (L) and two copies of a heavy chain (H) linked together by disulphide bridges. We can think of the molecule as a fork, with the handle and the base of the prongs consisting of constant regions and the tips of the prongs being variable. This variability is what allows the molecule to interact with the wide variety of antigens encountered during the life of the organism. The heavy chains consist of V (variable), D (diversity), J (joining) and C (constant) regions while the light chains have only V, J and C regions.

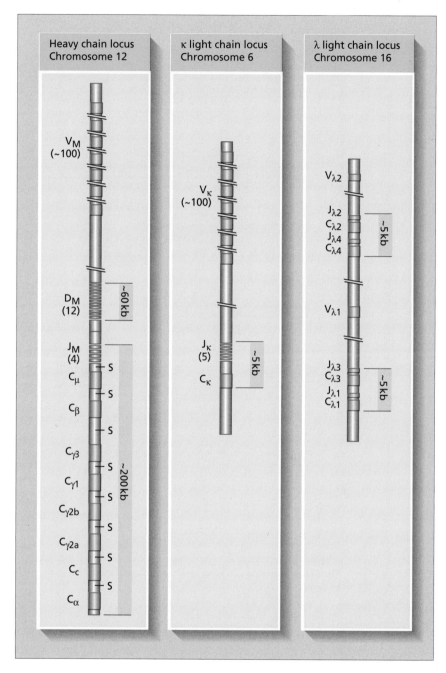

Fig. 6.9 Organization of the mouse immunoglobulin (Ig) genetic loci (after Engler & Storb, 1988). The mammalian Ig genes are clustered at three unlinked loci. In the mouse, the heavy chain genes are located on chromosome 12, while the light genes are located at two loci (the κ genes on chromosome 6 and the λ genes on chromosome 16). The approximate number of copies of different genes is shown. The heavy chain constant regions (Cμ–Cα) are the regions that can be interchanged by class-switching (see Section 6.6). Class-switching occurs at the switch regions denoted S.

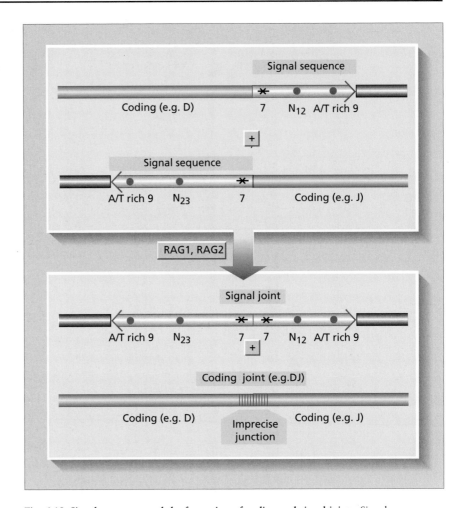

Fig. 6.10 Signal sequences and the formation of coding and signal joints. Signal sequences are composed of three parts. A 7 bp palindrome adjacent to the coding region, a spacer region of non-conserved nucleotides and a 9 bp AT-rich region. There are two types of signal sequence distinguished by the length of their spacer region of 12 or 23 bp. For a productive reaction to take place, there must be one signal sequence of each type present. The coding joint often includes additional bp at the junction between the two coding sequences.

recombining sequences. The coding joint by contrast is not precise and involves at least three types of variation. Some bases can be lost, some can be added by a mechanism that generates short inverted duplications of the coding sequence ends and nucleotides can be added by the enzyme-terminal transferase. These imprecise reactions are reminiscent of the incorporation of nucleotides that can occur during end-joining. Marty Gellert and his colleagues have demonstrated that, in mice that have a deficiency in the formation of coding joints, a high frequency of covalently closed hairpins can be detected. It is now known that these structures are normal intermediates in the reaction that are generated by the action of the proteins RAG1 and RAG2. Hairpin formation and cleavage generates the short inverted duplications often found at the coding joints (shown in Fig. 6.11).

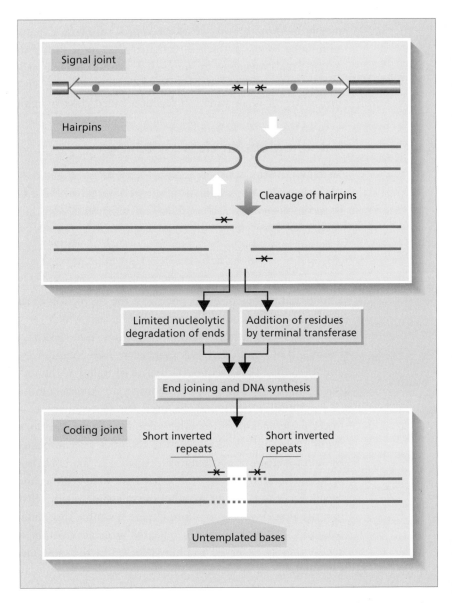

Fig. 6.11 Hairpin model for VDJ recombination. The formation of the coding joint involves first the formation of DNA hairpin structures that can then be cleaved to generate short inverted repeats. These ends are processed by exonucleolytic degradation the addition of residues by terminal transferase. Finally, the ends are joined in a reaction that sometimes but not always uses DNA homology between the single-stranded ends. This end-joining reaction may utilize the same enzymes involved in general end-joining.

When VDJ recombination has been completed the cell can, without further rearrangement, synthesize a μ heavy chain since the constant region Cμ is adjacent to the J segment. This stimulates the light chains to rearrange. They do this using the same mechanism and sequences as VDJ recombination. A complete IgM antibody has thus been made which will become membrane-bound and act as a receptor, allowing expansion of any clone stimulated by the antigen it recognizes. At some future stage there may be another rearrangement to change the constant part of the heavy chain while retaining the variable regions that recognize antigen. This is accomplished by a different recombination reaction called **class-switching.**

6.6   Immunoglobulin class-switching

Switching of the constant region of the heavy chain changes the function of the Ig without changing its ability to interact with a specific antigen. This is accomplished by a recombination reaction called class-switching. This occurs at long, internally repetitive, switch regions via a mechanism different to VDJ joining. These switch regions lie between the C_H genes, as shown in Fig. 6.9. Because they are located within intron sequences, recombination does not have to be very precise. The exact end-points of the recombination reactions do not affect the expression of the C_H genes themselves and recombination only determines which of the C_H genes is expressed. The switch regions vary from 1 to 10 kb in length and are composed of repetitive DNA sequences. Two common motifs are GAGCT and TGGGG. These sequences are often found at the recombination junction points that have been sequenced. Another motif is YAGGTTG (Y = pyrimidine) that may be used for recombination with a somewhat lower preference than the other repeats. The mechanism of recombination is unknown, but could involve the action of a recombinase. It may proceed by homologous recombination, however, the switch regions are not precisely homologous to each other over long regions and some switching events occur outside one or other of the switch regions involved. This could indicate that these events more closely resemble illegitimate recombination reactions. Deletions between repeated sequences within one switch region are commonly observed and may proceed by the same mechanism as class-switching.

6.7   Aberrant topoisomerase, replication and recombination reactions

Type I and type II topoisomerases are able to catalyse illegitimate events. Normally topoisomerase action involves strand-cleavage and religation to the same strands. This does not lead to recombination. However, occasionally, strand-transfer is accidentally catalysed. In *E. coli*, Hideo Ikeda has studied type II topoisomerase-mediated fusion of plasmid pBR322 and bacteriophage λ. This reaction can be catalysed by DNA gyrase and T4 topoisomerase. Type II topoisomerases catalyse double-strand-cleavage and strand-passage. It is thought that the illegitimate reaction involves the interaction of two gyrase cleavage sites and strand-transfer, instead of religation of parental strands. In eukaryotes, Mike Botchan has observed sequences resembling type I topoisomerase cleavage sites at the sites of excision of the SV40 provirus. This suggests that type I topoisomerases may also catalyse illegitimate events. Because type I topoisomerases catalyse single-strand-cleavage and religation, these events may involve more than one round of topoisomerase action or the processing of intermediates by other enzymes.

Proteins that interact with origins of replication or conjugal transfer can catalyse illegitimate recombination. Dusko Ehrlich has studied the illegitimate recombination reactions catalysed by nicking at the origin of replication of bacteriophage M13 and plasmid pC194, a plasmid that replicates via a single-stranded intermediate. In both these cases, deletions are formed where one end corresponds to the site of nicking at the origin of replication. Similar events are observed at the origin of transfer of the F factor plasmid.

Other illegitimate recombination reactions can occur when transposition or site-specific recombination involves the use of aberrant substrates. These include one-ended transposition, where one correct end of a transposon collaborates with a non-transposon sequence to transpose the intervening DNA. Another example of this class of event is the interaction of a site-specific recombination enzyme with an abnormal target sequence. This has been shown to occur in the cases of the bacteriophage λ integration and excision reactions and the G and C inversion systems of bacteriophages Mu and P1.

Another possibility is that illegitimate recombination is stimulated by a nearby homologous recombination reaction. This mechanism of promotion of illegitimate recombination has been proposed by Ichizo Kobayashi to explain the high frequency of illegitimate recombination associated with selection for homologous recombination between plasmids introduced into mammalian cells. In these studies, the end-points of illegitimate recombination reactions often coincided with the end-points of homology involved in the homologous interactions. Furthermore, similar illegitimate reactions were observed when these plasmid–plasmid homologous exchanges were selected in *E. coli* cells deficient in the genes involved in the resolution of Holliday junctions. His proposal is that certain cells (such as mammalian somatic cells or certain *E. coli* mutants) cannot resolve Holliday junctions efficiently and this can lead to strand-cleavage and illegitimate recombination.

Transposition of elements such as Ac and P can also lead to genome rearrangement by illegitimate recombination. It is proposed that transposition results in double-strand breaks which are normally repaired by end-joining or homologous recombination. However, if the two broken ends are not joined back together, they may join to other sequences, resulting in genome rearrangements which may involve multiple BFB cycles.

6.8   Illegitimate recombination and cancer

The development of many cancers is accompanied by extensive genome rearrangements. These rearrangements can precede tumorigenesis and are believed to be involved in the activation of oncogenes and the loss of tumour suppressor genes. In normal cells, check-points exist at the G_1–S and G_2–M points of the cell cycle that allow damaged chromosomes to be repaired before progressing to DNA replication (G_1–S) or cell division (G_2–M). In cells mutant for the p53 tumour suppressor gene, the G_1–S checkpoint is defective and this allows damaged chromosomes to enter S phase. This leads to BFB cycles and the accumulation of large amounts of amplified palindromic DNA (as described in Section 6.2). These amplification reactions can lead to further changes in gene expression (such as oncogene activation) and progression of tumour development. The p53 dependent G_1–S checkpoint is, for instance, missing in cells from individuals suffering from the inherited condition ataxia telangiecstasia.

In addition to rearrangements associated with BFB cycles, illegitimate recombination mediated by abnormal class-switching and VDJ recombination of Ig genes, can catalyse recombination between sequences not normally involved in Ig gene rearrangement. These events are frequently involved in the direct activation of oncogenes and are observed in lymphoid cancers such as B- and T-cell lymphomas and leukemias.

6.9   Cloning and targeting problems associated with illegitimate recombination

End-joining reactions are responsible for the problem faced by anyone wishing to target genes by homologous recombination in mammalian cells, as described in Chapter 7. Targeted recombination is 100–10 000 times less frequent than illegitimate recombination by end-joining which is extremely efficient. It appears that the chromosomes of higher eukaryotes have evolved a mechanism for the repair of strand-breaks that is simply to join them back together, irrespective of homology. This may be a sensible strategy for an organism that has a large proportion of intron DNA for which the end-points of joining reactions need not be precise. This may also be a sensible strategy for an organism with a large genome where the search for homology in somatic cells may be relatively difficult and where the danger of finding a chance homology located at another chromosomal location may be relatively high. If the DNA is normally well packed within chromatin, the broken ends of a chromosome may lie closely together and the best thing to do may simply be to join them together.

Another problem associated with illegitimate recombination is the cloning of eukaryotic DNA in bacteria. Some eukaryotic DNA sequences are deleted at a high frequency, by illegitimate recombination, in *E. coli* cells. Deletion is particularly favoured if the cloned DNA sequences are hard to replicate or enhance illegitimate recombination. The inhibition of replication associated with long palindromic sequences is alleviated in *sbcC* mutant hosts and a number of strains have been designed that combine *sbcC* with mutations that inhibit homologous recombination (e.g. *recA*) and DNA restriction (e.g. *hsdR, mcr, mrr*) to facilitate the cloning of eukaryotic DNA sequences.

6.10   Conclusion

We have seen that illegitimate recombination reactions can be subdivided into two primary mechanisms: end-joining and strand-slippage. In addition to these primary mechanisms, there are many events that are better described as aberrant forms of other reactions, mediated by the enzymes of DNA replication, topoisomerases and recombination proteins themselves. These aberrant reactions may be self-contained or involve end-joining or strand-slippage as part of an event initiated by some other reaction.

Illegitimate recombination produces a multitude of chromosomal rearrangements. Many of these result in unviable or deleterious phenotypes. Some result in the deregulation of cell growth and carcinogenesis. Perhaps some escape these two fates and lead to novel chromosomal organizations, therefore providing the most flexible raw material for the reorganization and evolution of the genome.

The recombination reactions involved in Ig gene rearrangement have similarities to illegitimate recombination, although they are focused to specific regions. The difficulty in classifying these reactions highlights the fact that the subdivision of recombination into homologous, site-specific, transpositional and illegitimate events is only a logical classification to aid our understanding. As we continue to understand more about the mechanisms of individual recombination reactions, we will begin to break down the barriers between these subdivisions.

Further reading

Allgood N. D. & Silhavy T. J. (1988). Illegitimate recombination in bacteria. In Kulcherlapati R. & Smith G. R. (eds) *Genetic Recombination*, pp. 309–330. American Society for Microbiology, Washington.

Champoux J. J. & Bullock P. A. (1988). Possible role for eukaryotic type I topoisomerase in illegitimate recombination. In Kulcherlapati R. & Smith G. R. (eds) *Genetic Recombination*, pp. 655–667. American Society for Microbiology, Washington.

Ehrlich S. D. (1989). Illegitimate recombination in bacteria. In Berg D. E. & Howe M. M. (eds) *Mobile DNA*, pp. 799–832. American Society for Microbiology, Washington.

Ehrlich S. D. *et al.* (1993). Mechanisms of illegitimate recombination. *Gene* **135**, 161–166.

Engler P. & Storb U. (1988). Immunoglobulin gene rearrangement. In Kulcherlapati R. & Smith G. R. (eds) *Genetic Recombination*, pp. 667–700. American Society for Microbiology, Washington.

Meuth M. (1989). Illegitimate recombination in mammalian cells. In Berg D. E. & Howe M. M. (eds) *Mobile DNA*, pp. 833–860. American Society for Microbiology, Washington.

Leach D. R. F. (1995). Cloning and characterization of DNAs with palindromic sequences. In: Setlow J. (ed.) *Genetic Engineering, Principles and Methods*. Plenum Publishing, London.

Leach D. R. F. (1994). Long DNA palindromes, cruciform structures, genetic instability and secondary structure repair. *BioEssays* **16**, 893–900.

Lewis S. & Gellert M. (1989). The mechanism of antigen receptor gene assembly. *Cell* **59**, 585–588.

Lieber M. R. (1992). The mechanism of V(D)J recombination: A balance of diversity, specificity, and stability. *Cell* **70**, 873–876.

Lutzker S. G. & Alt F. W. (1989). Immunoglobulin heavy-chain class switching. In Berg D. E. & Howe M. M. (eds) *Mobile DNA*, pp. 693–714. American Society for Microbiology, Washington.

Roth D. & Wilson, J. (1988). Illegitimate recombination in mammalian cells. In Kulcherlapati R. & Smith G. R. (eds) *Genetic Recombination*, pp. 621–654. American Society for Microbiology, Washington.

Schatz D. G., Oettinger M. A. & Schlissel M. S. (1992). V(D)J recombination; molecular biology and regulation. *Ann. Rev. Immunol.* **10**, 359–383.

Toledo F., Buttin G. & Debatisse M. (1993). The origin of chromosome rearrangements at early stages of *AMPD2* gene amplification in Chinese hamster cells. *Current Biology* **3**, 255–264.

References

Darlow J. M. & Leach D. R. F. (1995). The effects of trinucleotide repeats found in human inherited disorders on palindrome inevitability in *Escherichia coli* suggest hairpin folding preferences *in vivo*. *Genetics* **141**, 32–40.

Engler P. & Storb U. (1988). Immunoglobulin gene rearrangement. In Kulcherlapati R. & Smith G. R. (eds) *Genetic Recombination*, pp. 667–700. American Society for Microbiology, Washington.

Trinh T. Q. & Sinden R. R. (1991). Preferential DNA secondary structure mutagenesis in the lagging strand of DNA replication in *E. coli*. *Nature* **352**, 544–547.

7 Applications of Genetic Recombination

7.1 Introduction

Historically, genetic recombination has been used for two purposes. One is the mapping of genetic loci on chromosomes and the second is the construction of strains. These remain the major applications. However, the advent of somatic gene therapy creates a third.

The genetic mapping of chromosomes from bacteriophages to humans has revolutionized biology. It has provided the information with which we logically order genes and understand their organization and regulation. This knowledge has allowed us to use genetic loci as markers for other genes located nearby, thus facilitating strain construction, cloning and eventually sequencing. Marker loci have also provided reference points for the tracking of medically important genes in affected families.

The construction of strains is at the centre of the genetic approach to biology and is one of the cornerstones of modern biology. Strain construction is the surgeon's approach to chromosomes. We can ask: What happens if we remove one amino-acid from a protein or replace it with another? What happens if we make too much of one protein or none at all? What happens if we make a defective protein in the presence of its active counterpart? What happens when we make specific changes to regulatory circuits? In conjunction with biochemistry, we can ask: What are the *in vivo* consequences of the activities of proteins analysed *in vitro*? We can also identify genes encoding proteins that can then be studied *in vitro*.

The tools of the genetical surgeon are mutagenesis and recombination. These are not mutually exclusive since recombination can be used for mutagenesis. For example, transposition can be used to obtain insertion mutants and illegitimate recombination can be used to isolate deletion mutants. However, generally speaking, strain construction requires the isolation of mutants and then their manipulation into the desired genetic backgrounds by recombination. The advent of techniques of *in vitro* genetic manipulation with restriction enzymes has further blurred the lines between mutagenesis and recombination. The cutting and pasting reactions performed by the genetic engineer are recombination reactions of a kind and they are often also performed for the purpose of mutagenesis. Since *in vitro* recombination using restriction enzymes is not a natural process it is not covered in this book. Included in this chapter are some of its uses when coupled to 'natural recombination'.

7.2   Mapping in eukaryotes

Thomas Hunt Morgan's proposal that linkage and recombination reflected the relative location of genes on chromosomes dates back to 1911. During the next two decades, he, his three students, H. J. Muller, A. H. Sturtevant and C. B. Bridges and their colleagues carried out many experiments that clarified the linkage relationships between genes and led to the development of **linkage maps** of *Drosophila* chromosomes.

Mapping is carried out by performing crosses between mutants and measuring the recombination frequencies amongst the progeny. The closer two mutations are located along the length of the chromosome, the lower is the recombination frequency. If many crosses are performed, a map of the relative positions of genetic loci can be obtained. This seems simple enough. However, for a map to be really useful, it is important to obtain information both on relative position and distance. These distances should be additive to enable direct comparisons of the relative positions of different loci. Unfortunately recombination frequencies are not always additive and must therefore be converted into units that are. These are known as **map units** or **centimorgans**. In order to illustrate this, the derivation of a simple mapping function for a diploid organism is described below.

Imagine a pair of homologous chromosomes of a hypothetical diploid organism undergoing meiosis. Furthermore, let's impose a rule that one and only one crossover is permitted along the whole length of the chromosome per meiosis. We can then construct a linkage map by observing the recombination frequencies between marked loci. We can define a recombinational distance (d) between any two marked loci as being equal to the probability of crossing over (R) between these loci. Or: $R = d$.

Recombination frequencies will be additive throughout the chromosome and there will be a linear relationship between recombination frequency and map units as shown in Fig. 7.1. Because, to a first approximation, recombination occurs after DNA replication, the frequency of crossing over (R) reaches a maximum of 50% and is equal to half the frequency of a chiasma forming between any two marker loci. The factor of one half comes about because only two chromatids have undergone crossing over in one chiasma. The other two chromatids remain parental,

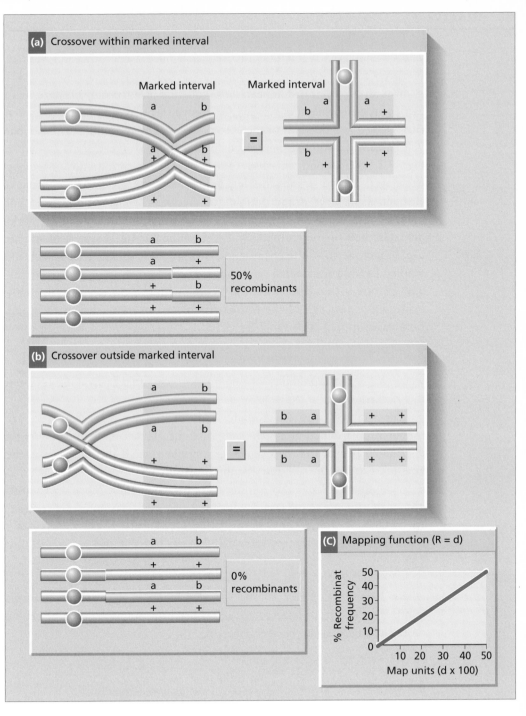

Fig. 7.1 **Mapping in a hypothetical chromosome where one and only one crossover occurs per meiosis.** In this hypothetical chromosome, crossing over must either occur within the marked interval or outside it. When it occurs within the interval, 50% of the products are recombinant (a), while if it occurs outside the interval, none of the products are recombinant for the marked loci (b). This leads to a straightforward mapping function of $R = d$ which reaches a maximum of 50% for loci at the extremities of the chromosome (c).

as shown in Fig. 7.1. If we call the frequency of chiasma formation X, then: $R = \frac{1}{2} X$.

We can ask what will happen if the rule that one and only one crossover can occur per bivalent is relaxed. As above, if a chiasma occurs between two marked loci one half of the chromatids will be recombinant. But what will happen if two chiasmata occur in this marked interval? As shown in Fig. 7.2, the second chiasma can involve both chromatids originally exchanged (a **two-strand double**), only one of the previously exchanged chromatids (a **three-strand double**), or the two chromatids that were not involved in the previous exchange (a **four-strand double**). Again, remarkably, one half of the chromatids are recombinant for the marked loci. It can, in fact, be shown that any number of multiple exchanges between the marked loci always results, on average, in the formation of 50% recombinant chromosomes. Therefore, the recombination frequency can be simply expressed as one half of the probability of *one or more* chiasmata occurring between the marked loci, or:

$$R = \frac{1}{2} (X^{1+})$$

where X^{1+} is the probability of one or more chiasmata occurring between the marked loci.

The simplest way to estimate X^+ is to assume that chiasmata are randomly distributed along the length of the chromosome and randomly distributed between chromosomes in a population. If this is the case, the Poisson distribution will describe the probabilities of formation of 0, 1, 2, 3 . . . chiasmata in any marked interval in the population. Since we want to obtain an estimate of the events where one or more chiasmata have formed, we calculate the frequency of events where no chiasmata have occurred (X^0) and subtract this from one. From the Poisson distribution:

$$X^0 = e^{-Xm}$$

where Xm is the mean number of chiasmata in the marked interval. Subtracting this from one, we get an estimate for the probability of one or more chiasmata in the marked interval.

$$X^{1+} = 1 - e^{-Xm} \qquad \text{and therefore:}$$

$$R = \frac{1}{2} (1 - e^{-Xm})$$

Since one chiasma involves two of the four chromatids in a bivalent, the mean number of chiasmata in a marked interval is two times the recombinational distance: Xm = 2d, and:

$$R = \frac{1}{2} (1 - e^{-2d})$$

This is the **mapping function** derived by J. B. S. Haldane and known as the **Haldane function**. This results in a non-linear relationship between R and d (shown in Fig. 7.2(b)). The larger the interval, the higher the probability that two or more chiasmata may be present in the population of recombining chromosomes. Since events that involve multiple chiasmata result, on average, in 50% recombination and events with no chiasmata result in 0% recombination, the recombination frequency is simply reduced by the proportion of events that have no chiasmata in the marked interval. Figure 7.2(b) also shows that real data, collected in three different organisms, more closely resemble a linear function at low recombination frequencies and a Haldane function at high frequencies. In the real world, chiasmata are not distributed randomly according to the Poisson distribution and interference (described in Chapter 2) leads to a close approximation of the data to a linear relationship at recombination frequencies below 20%. This does allow quite accurate maps to be constructed without complex mapping functions if relatively closely linked markers are used. In the range of 1–20% recombination, one **map unit** or (**centimorgan**) approximates a 1% recombination frequency.

Note that d is not a measure of real physical distance but only of the probability of recombination. There is not a predictable relationship between recombinational distance (measured in map units) and physical distance (measured in base-pairs). The physical distance corresponding to 50 map units is simply the distance where an average of one crossover occurs per bivalent. Furthermore, if the probability of a crossover is not uniformly proportional to physical distance along the length of a chromosome, there will be no correlation between map units and distance. This situation is likely to obtain in any real chromosome, where recombinators will stimulate exchanges in particular regions and other factors, such as inversions, will reduce or abolish recombination in others.

Two and three factor crosses have led to the construction of genetic maps in a wide range of experimental organisms from bacteriophages to mammals. The appropriate mapping function will depend on the nature of the experimental system. The knowledge of how recombination frequencies can be related to genetic distances in map units by the application of the appropriate mapping function has been essential and readers are referred to Stahl (1979) for a more advanced treatment of this subject.

An example of a three-factor cross between *Drosophila* mutants is shown in Fig. 7.3. Three-factor crosses are informative since they can be used to order three marked loci without information on genetic distances.

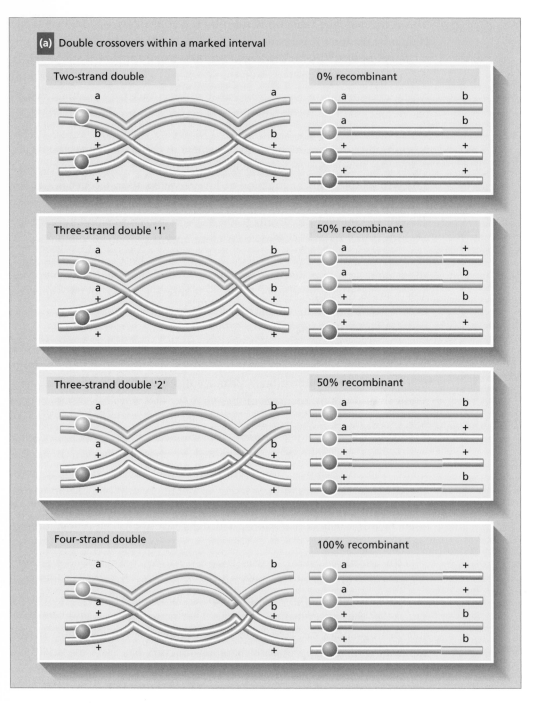

Fig. 7.2 Mapping in a chromosome where multiple crossovers are permitted. (a) The four different types of double crossover that can occur between two marked loci are shown. In the two-strand double, none of the products are recombinant; in both classes of three-

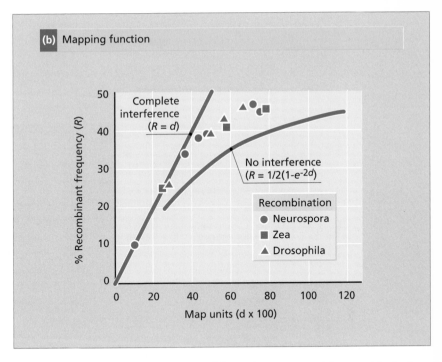

Fig. 7.2 (*Continued*) strand double, 50% of the products are recombinant; and in the four-strand doubles, all the products are recombinant. If each of these four types of double crossover is equally likely to occur, the proportion of recombinant products is 50%. (b) This leads to the proposal of the Haldane function which is shown compared to real data obtained from *Drosophila*, maize and *Neurospora*, after Perkins (1962).

We can see from this example that two of the six possible classes of recombinant chromosomes are present at a much lower frequency than the other two. This is because (except for events involving very close markers, as discussed in Chapter 2) two of the six possible classes of recombinants require double crossovers, whereas four require only a single crossover. Without knowing the order of the genes it is immediately clear that *vermilion* lies between the other two.

Strain construction in organisms that reproduce sexually has classically been achieved by crossing and selecting progeny with the desired genotype. This is a powerful method that relies on homologous recombination between genes and independent assortment of genes on different chromosomes. However, the precise and directed modification of small regions of eukaryotic genomes has, in the past, been difficult because of the lack of gene-transfer systems. This has now changed with the advent of gene-targeting methods (described in Section 7.6).

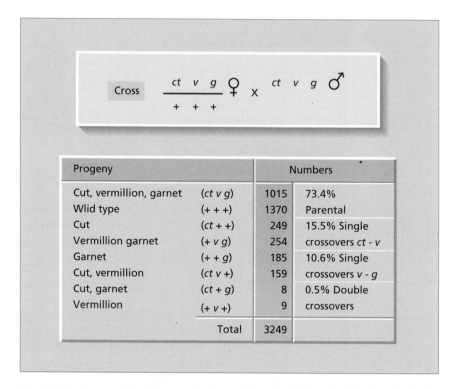

Fig. 7.3 **Data from a three-point cross in *Drosophila*,** after Bridges & Olbrycht (1926). The cross shown is between flies carrying the marked loci cut (*ct*), vermilion (*v*) and garnet (*gt*) on their X chromosome. *ct* confers cut wings, *v* bright scarlet eyes and *gt* purple eye colour. The data for male and female progeny has been taken together and presented as the numbers of parental and recombinant chromosomes detected. It can be seen that two of the six possible classes of recombinants are much rarer than the other four. This immediately reveals that *v* must lie between the other two loci.

7.3　Linkage analysis in humans

The analysis of linkage in humans is of fundamental importance for three reasons. Firstly it is used to help track, in family trees, the genes responsible for genetic disease. Information on linkage is necessary when markers for the disease causing alleles themselves are not available. This contributes to accurate diagnosis and allows genetic counselling of individuals and their families. Secondly it has allowed the positional cloning of human genes such as that encoding the cystic fibrosis transmembrane conductance regulator protein (CFTR). Linkage analysis was used to locate the CFTR gene relative to other markers and this information allowed the isolation of the gene itself. And thirdly it is essential for the construction of a genetic map of the human genome. It is therefore contributing to the genome mapping projects which aim to map and eventually sequence all the approximately 80 000 human genes.

There is, however, a major problem in studying human linkage. Recombinant frequencies obviously cannot be obtained by setting up crosses. In the example given in Fig. 7.3, a total of 3249 flies were examined to obtain accurate recombinant frequencies. Instead, human recombinant frequencies must be inferred from pedigree analysis of families, wherever the relevant information is available. This usually means dealing with small samples and with incomplete sets of data.

What is required is not only a method of measuring linkage by studying family trees but also a way of determining the statistical significance of any result obtained. The solution that is generally used is to compare the probability of finding the observed frequencies of recombinants given linkage and a particular frequency of recombinant to that predicted for no linkage and maximum recombination due to independent assortment. The measure obtained is thus an **odds ratio** or an **odds for linkage** at a particular recombinant frequency.

In this analysis, the recombinant frequency (which is often referred to as the **recombinant fraction** because of the small data set available) is normally denoted θ. θ varies between 0, for completely linked loci, and 0.5, for unlinked loci. This is because unlinked loci will recombine with each other 50% of the time by random segregation. For example, in a cross, AB/ab×ab/ab, between unlinked genes a 1:1:1:1 ratio of progeny with genotypes AB/ab, Ab/ab, aB/ab and ab/ab is predicted by Mendelian segregation. Two of these are parental and two are recombinant (50% recombination).

The odds ratio is defined as:

$$\text{Odds ratio} = \frac{\text{Probability of occurrence of all the phenotypes assuming recombinant fraction } \theta}{\text{Probability of occurrence of all the phenotypes assuming recombinant fraction } 0.5}$$

or:

$$L^*(\theta) = \frac{L(\theta)}{L(0.5)}$$

In order to facilitate the mathematical handling of these values, it has been convenient to use the logarithm of $L^*(\theta)$ or the **logarithm** of the **odds ratio**. This value, denoted $Z(\theta)$, is what has become known as the **lod** score:

$$Z(\theta) = \log_{10} L^*(\theta)$$

$Z(\theta)$ is thus a measure of the relative likelihood of linkage at a particular recombinant fraction (θ) to the likelihood of no linkage. It must therefore be estimated at different values of θ to determine a maximum lod score.

This maximum lod score (Z_{max}) provides two pieces of information. Its value is a measure of the likelihood of linkage and the value of θ at which it reaches a maximum is an approximate measure of linkage. This is illustrated in Fig. 7.4 which shows an example of the relationship between the lod score and recombinant fraction. In practice the arbitrary value of $Z_{max} > 3$ is taken as good evidence for linkage. This represents a 1000:1 odds ratio in favour of linkage.

In general, the estimation of Z(θ) is very complex because of incomplete information and various complications such as differences between recombinant frequencies in males and females. Therefore, it is normally estimated by computer using specially devised programmes. However, the values obtained are simple logarithms of odds ratios and are therefore additive. This has the advantage that data obtained on the same loci in separate studies can simply be summed to obtain larger values of Z(θ) and more accurate estimates of θ. In simple situations Z(θ) can be estimated without the help of the computer and an example is given in Fig. 7.5 to illustrate several features of this type of analysis.

If a gene has been cloned and mutant alleles identified, these can be followed directly in individuals and their family members by using specific probes. An example of such an approach is the tracking of the ΔF_{508} mutation which is the cause of 50–80% of the incidence of cystic fibrosis. However, what can one do for the rare alleles of cystic fibrosis or for genes that have not been cloned? In these situations one must use indirect tracking using information on linkage. If a disease-causing locus has been mapped relative to another polymorphic locus and shown to be very closely linked,

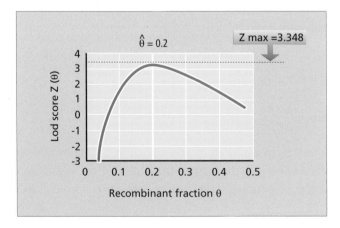

Fig. 7.4 The relation between lod score and recombinant fraction. After Ott, 1991. A hypothetical example is presented for eight recombinants out of 40 individuals. θ the actual fraction of recombinants is 8/40 = 0.2 and this value of θ gives the highest value of Z(θ) = 3.348. At other values of θ, Z(θ) is lower giving rise to the curve shown.

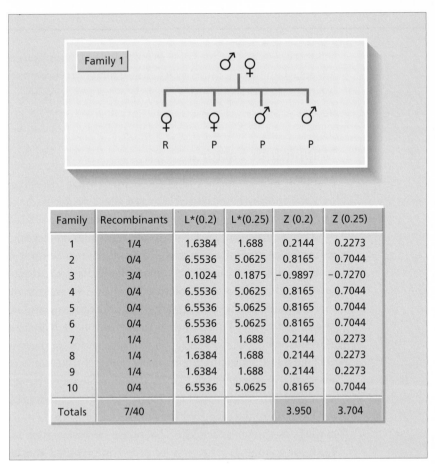

Family	Recombinants	L*(0.2)	L*(0.25)	Z (0.2)	Z (0.25)
1	1/4	1.6384	1.688	0.2144	0.2273
2	0/4	6.5536	5.0625	0.8165	0.7044
3	3/4	0.1024	0.1875	−0.9897	−0.7270
4	0/4	6.5536	5.0625	0.8165	0.7044
5	0/4	6.5536	5.0625	0.8165	0.7044
6	0/4	6.5536	5.0625	0.8165	0.7044
7	1/4	1.6384	1.688	0.2144	0.2273
8	1/4	1.6384	1.688	0.2144	0.2273
9	1/4	1.6384	1.688	0.2144	0.2273
10	0/4	6.5536	5.0625	0.8165	0.7044
Totals	7/40			3.950	3.704

Fig. 7.5 A simple example of the application of lod scores. Imagine that two loci exist that are linked and, given an infinite sample, would show a recombinant frequency of 0.2. Furthermore let's imagine that, as in Fig. 7.4, our actual sample size is only 40 and is divided into 10 families with four children each. The first family that is studied has two girls and two boys and is shown in this figure. One of the girls is recombinant (R) while the other children are parental (P). With hindsight, we know that the real probability of being recombinant is 0.2 and of being parental is 0.8. However, in this first family the recombinant fraction is 0.25. Let's calculate $Z(\theta)$ for both these recombinant fractions. In this simple example $L(\theta) = \theta(1 - \theta)^3$ (the product of the probabilities of each of the individuals).

For $\theta = 0.2$, $L(0.2) = 0.2(0.8)^3 = 0.1024$
For $\theta = 0.25$, $L(0.25) = 0.25(0.75)^3 = 0.1055$
For $\theta = 0.5$, $L(0.5) = 0.5(0.5)^3 = 0.0625$

Therefore:

$Z(0.2) = \log_{10}(L(0.2)/L(0.5)) = 0.2144$
$Z(0.25) = \log_{10}(L(0.25)/L(0.5)) = 0.2273$

As we would expect, $Z(0.25)$ is higher than $Z(0.2)$ because the fraction of recombinants is 25% but the sample is so small that little confidence can be given to the estimate of $\theta = 0.25$. This is manifested in the low value of $Z_{max} = 0.227$. In family 2, there are in fact no recombinants and the numbers of recombinants in the rest of the 10 families is shown in the table. We can see that out of the total of 40 individuals 7 are recombinants and that when summed over all the families, $Z(0.2) = 3.950$ is higher than $Z(0.25) = 3.704$. In fact the observed fraction of recombination was $\frac{7}{40}$ or 0.175. $L(0.175) = (0.175)^7 \times (0.825)^{33}$ and $Z(0.175)$ has a value of 3.986. This is the observed Z_{max}. What can be concluded from this hypothetical example is that the genes under investigation are linked and that their predicted recombination frequency is approximately 0.2.

this other locus can act as a marker for tracking the gene of interest in affected families. Because of its close linkage, recombination is unlikely to separate the two loci and to a close approximation they segregate as one unit.

7.4 Mapping in bacteria

Mapping in bacteria is complicated by the fact that they are haploid and do not reproduce sexually. However, these problems have been overcome by harnessing mechanisms of DNA transfer.

The mechanism that most closely resembles eukaryotic crosses is **conjugation** (also known as bacterial mating). Here a **donor** (F⁺) cell is mated with a **recipient** (F⁻) and occasionally DNA from the donor chromosome is transferred into the recipient whereupon it can recombine. The F plasmid that is present in F⁺ cells is responsible for the mating and occasionally transfers chromosomal DNA when it happens to have integrated into the donor chromosome. This is a very infrequent event and strains where the F plasmid has already integrated in the donor chromosome have been isolated that transfer chromosomal DNA at a much higher frequency. These strains are called **Hfr** strains for high frequency of recombination. Long stretches of DNA (up to the length of the whole chromosome) are transferred in such matings as shown in Fig. 7.6. There are two ways in which conjugation is used for mapping. The first is called **interrupted mating**. Here, mating between an Hfr donor and a recipient is initiated and then interrupted at various times by vigorous vortexing, or the use of antibiotics that inhibit DNA replication. Markers close to one side of the site of integration of the F plasmid are transferred early and more distant markers are transferred later. Markers can, therefore, be ordered by their time of entry to the recipient cell. The second method is to measure recombination frequencies between markers. Usually here, a marker that is transferred late is selected and the frequency of co-inheritance of an earlier marker is scored. The

Fig. 7.6 **Bacterial conjugation (mating).** F⁺ strains carry the plasmid F that is transmissible from donor to recipient cell. F carries genes that are required for conjugation. These include genes for the synthesis of **pili** that are responsible for bringing mating pairs together. Occasionally an F plasmid becomes integrated into the host chromosome by homologous recombination. Strains that carry an integrated F plasmid (Hfr strains) transfer chromosomal DNA at a high frequency to F⁻ recipients. The DNA of F is nicked at the origin of transfer (*oriT*) and a 5′ end is passed from donor to recipient. As transfer proceeds, the strand is copied to regenerate double-stranded DNA by rolling-circle replication. The DNA can then recombine with the resident chromosome. In this figure, the donor chromosome is marked with a wild-type gene (+) and the recipient carries a mutation in this gene (–).

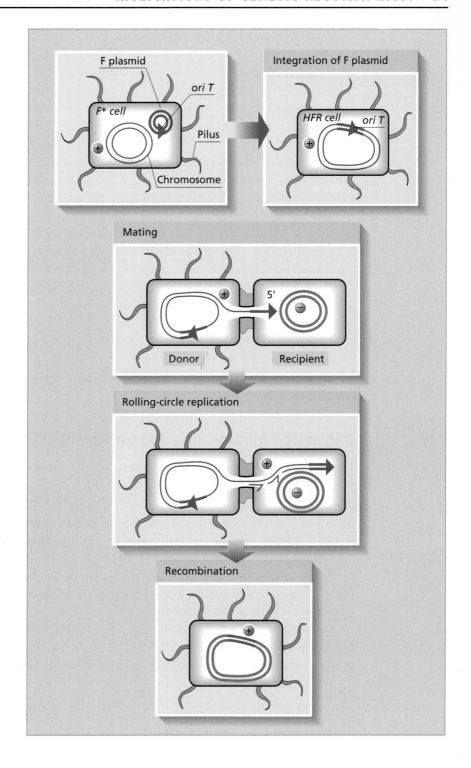

closer the markers are linked, the higher the co-inheritance. These methods are shown in Fig. 7.7.

Conjugation is useful for the rough mapping of mutations but is less useful for the mapping of closely linked genes. For fine mapping, recombination frequencies are usually measured after introduction of DNA by **generalized transduction.** Generalized transduction requires the packaging of pieces of chromosomal DNA into bacteriophage heads. This DNA can then be delivered to a recipient strain where it can recombine with the host chromosome. Bacteriophages, such as P22 of *Salmonella typhimurium* or P1 of *Escherichia coli,* mistakenly package chromosomal DNA sequences at a low frequency and therefore mediate generalized transduction as shown in Fig. 7.8. Because the amount of DNA packaged into a bacteriophage head is small (e.g. 100 kb for phage P1) only markers lying within this distance can be mapped by co-transduction as shown in Fig. 7.9.

For most purposes of strain construction, a modification to a small region of DNA is required and it is desirable to retain the original genotype of the rest of the chromosome. To do this, it is necessary to introduce only a limited amount of donor DNA. Therefore, a very useful method for bacterial strain construction is generalized transduction. For even more precise alterations of bacterial chromosomes, gene-targeting can be used (see Section 7.6).

7.5   Transposon mutagenesis and transposable vectors

Transposons have been widely used for the mutagenesis of both prokaryotic and eukaryotic genomes. The insertion of a transposable element will result in gene inactivation and concurrently tag that gene with any markers carried by the transposon. These may for instance be drug-resistance genes or reporter genes such as β-galactosidase.

The presence of the transposable element can also facilitate the cloning of the gene into which it is inserted. This can either be because the transposon is engineered to encode a selectable marker or be used as a target for a transposon specific DNA probe.

Many variations upon this scheme are possible. Vectors carrying a reporter gene without its promoter and/or ribosome-binding site can be used to select gene and operon fusions and eukaryotic vectors lacking an

---

Fig. 7.7 Mapping by conjugation. (a) **Interrupted mating.** Donor and recipient cells are mixed and samples removed after different time periods. The samples are vortexed rapidly to disrupt the mating pairs and are plated on media selective for markers a^+, b^+ or c^+. At 10 minutes after initiation of transfer, marker a^+ begins to appear, at 17 minutes marker b^+ and at 25 minutes marker c^+. It can be concluded that markers a^+, b^+ and c^+ are arranged in that order and that a^+ is the closest to the origin of transfer (*oriT*) of F. Because the rate

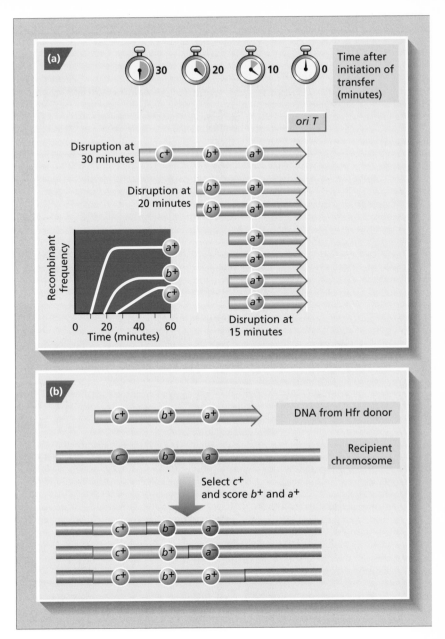

Fig. 7.7 (*Continued*) of DNA transfer in conjugation is constant, the times of entry correspond linearly to the distance from *oriT*. Mating pairs have a significant rate of spontaneous disruption (in the absence of any artificial vortexing) which causes the curves to plateau at different recombinant frequencies but does not affect the earliest time of entry. (b) **Co-inheritance of markers.** Donor and recipient cells are mixed and mating allowed to proceed. The cells are then plated out on a medium selective for recipients that have acquired marker *c⁺*. This ensures that all recombinants analysed have received the DNA between *oriT* of F and *c⁺*. The recombinants are then analysed for the coinheritance of markers *a⁺* and *b⁺*. A genotype *c⁺b⁻* is obtained by recombination between *c* and *b*. Similarly a genotype *b⁺a⁻* is obtained by recombination between *b* and *a*. The recombination frequencies are a measure of the recombinational distance between the markers.

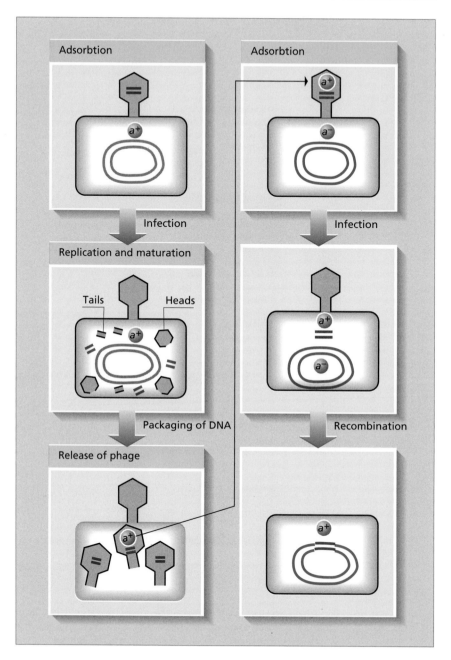

Fig. 7.8 Generalized transduction. A bacteriophage such as P1 of *E. coli* or P22 of *S. typhimurium* adsorbs to a bacterium. Its DNA is injected into the host and begins to replicate. Bacteriophage components such as heads and tails are made and the DNA is packaged into new particles. Occasionally, host chromosomal DNA is packaged instead, to generate transducing particles. If a lysate grown on a culture of a host of genotype a^+ is used to infect a recipient of genotype a^-, transducing particles containing a^+ DNA will be injected into a small number of target cells. These cells can then generate a^+ transductants by homologous recombination

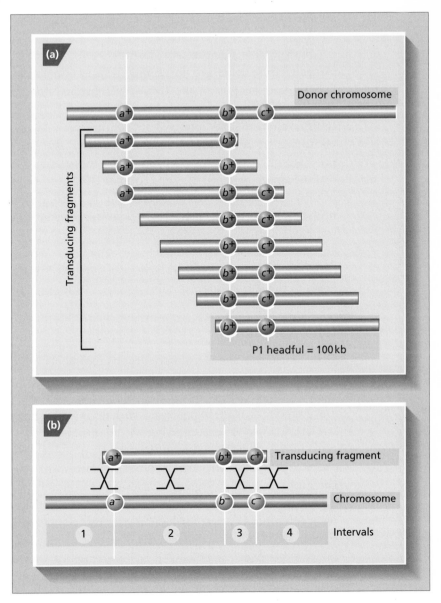

Fig. 7.9 Mapping by co-transduction. For the fine mapping of close markers, co-transduction provides the method of choice. Co-transduction frequencies are a measure of the separation of markers for two reasons: packaging and recombination. (a) **Packaging.** Bacteriophages such as P1 or P22 that mediate generalized transduction package a headful of DNA. For P1 this is 100 kb long. The likelihood of two markers being co-packaged within one phage head will depend on their proximity on the chromosome. In the illustration, markers a^+ and b^+ are co-packaged three times, a^+ and c^+ once, and b^+ and c^+ six times. (b) **Recombination.** For any transducing fragment, the probability of co-transduction of two markers will depend on co-inheritance at the stage of recombination. In the illustration, there are three markers on the transducing fragment and four intervals (marked 1–4) that can be used to integrate the DNA. At this stage, close markers are more likely to be co-inherited than distant markers.

enhancer can be used to obtain insertions close to a chromosomal-enhancer sequence.

Transposable elements also provide a mechanism for the delivery of cloned DNA sequences to the genome. This has been most successfully used for the manipulation of the *Drosophila* genome using derivatives of the transposable element known as P (see Chapter 5). P is a transposon that transposes via a DNA intermediate and has been engineered to allow the insertion of foreign DNA adjacent to a selectable marker. Transposase, provided in *trans* allows insertion of the transposon. Germ-line transformation can be obtained after injection of DNA into embryos. An example of such a transposable vector is shown in Fig. 7.14. A similar approach has been developed for the delivery of cloned sequences to human cells using retroviral vectors. Retroviruses transpose a DNA copy of their RNA genome to the host chromosome (see Chapter 5, Fig. 5.14). Genetically manipulated retroviral vectors have been developed that can deliver cloned genes to the chromosome in this way (Fig. 7.11).

7.6   Gene-targeting

Gene-targeting is the use of homologous recombination to modify a chromosomal gene with a cloned copy. This technique is very powerful because it combines the power of genetic engineering with strain construction to make precise changes to a genome without introducing other alterations.

The first use of gene-targeting was to manipulate *E. coli* chromosomal genes using cloned sequences. Bacteriophage λ vectors are particularly useful for this since they can be reversibly integrated and excised from the chromosome by homologous recombination. This strategy is shown in Fig. 7.12. The power of gene-targeting was not fully realized until techniques were developed to select for homologous recombination after introduction of linearized DNA directly into cells. This type of experiment was pioneered in *Saccharomyces cerevisiae* and has been applied with dramatic effect to the mouse. In the early 1980s, experiments were performed by Jack Szostak and colleagues that revolutionized yeast genetics. They showed that DNA introduced into *S. cerevisiae* could transform cells by homologous recombination with chromosomal genes and that this recombination was stimulated by cleavage of the introduced DNA within the region of homology. They also showed that if a gap was made in the incoming DNA it was repaired by copying the chromosomal information. These observations led to the proposal of the double-strand break-repair model of recombination described in Chapter 2. The revolutionary nature of these experiments was that they opened the way to introducing genetically manipulated genes into the genome of yeast and to similar experiments in cultured mammalian

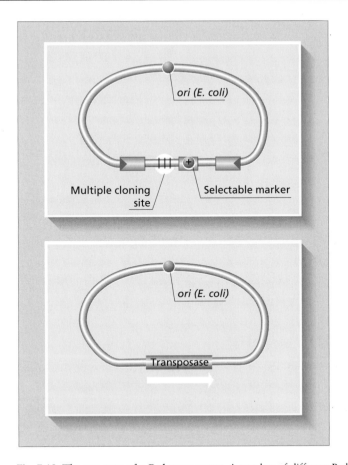

Fig. 7.10 The structure of a P element vector. A number of different P element vectors have been developed that contain different components. However, the most commonly used vector design is that shown. A circular plasmid containing an origin of replication for *E. coli* carries a defective P element that includes the DNA sequences required for transposition but not the transposase gene. Between the inverted repeats of the element there is a multiple cloning site into which DNA fragments can be inserted and a gene that can be used to select for the transposon (designated +). This is often the *rosy* (*ry*⁺) gene that can be identified by complementa'tion of *ry*⁻ eye colour or by selection on purine-containing medium. Transposase is usually supplied in *trans*, either on a plasmid encoding the gene but without terminal inverted repeats, or by co-injection of purified transposase protein.

cells. In *E. coli* these strategies can be used in *recD* mutant hosts that have an altered RecBCD enzyme that is competent for recombination but does not degrade linear DNA. A summary of the four basic strategies for gene-targeting with linearized DNA is shown in Fig. 7.13.

The first experiments demonstrating gene-targeting in mammalian cells were carried out using genes for which there was a strong selection for homologous recombinants. This is because illegitimate recombination by end-joining is between 1000 to 10 000 times more frequent than homologous

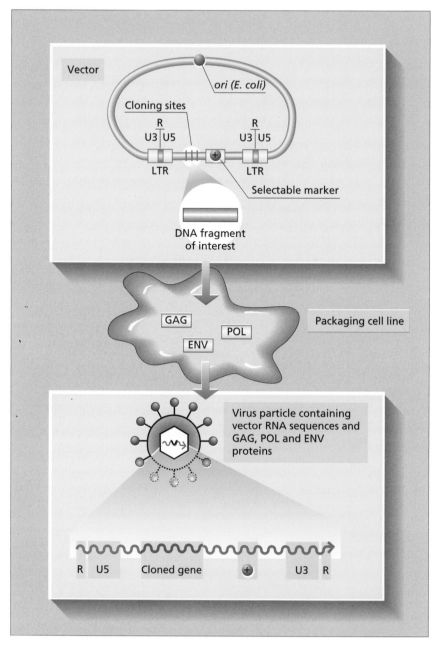

Fig. 7.11 The structure of a retroviral vector. Retroviral vectors consist of the LTR sequences of a retrovirus flanking DNA to be transposed. This often contains a genetic marker that confers a selectable property (e.g. neomycin resistance) and sites that are useful for the insertion of DNA fragments of interest. Once the DNA of interest has been inserted in the vector, the plasmid DNA is introduced into a packaging cell-line that contains one or more plasmids expressing the *gag*, *pol* and *env* genes. Their gene products allow the packaging of the vector RNA into a virus particle that also contains the gene product required for the reverse-transcription and integration of the vector in the target cells. Once integrated, the DNA is stable since the vector is devoid of viral genes.

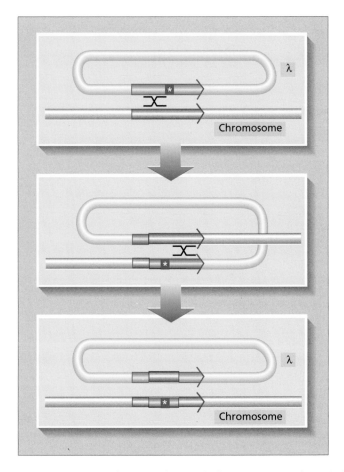

Fig. 7.12 Targeting of genes in the *E. coli* chromosome using bacteriophage λ vectors. The gene of interest is first cloned in an integration-deficient λ vector. The DNA is then altered in the desired way, a new host is infected and lysogens are selected. Since the λ vector is integration deficient, it will only be able to form a prophage by recombination with the host chromosome via the region of cloned homologous DNA. All that remains is to select for loss of the prophage by recombination between the directly repeated copies of chromosomal DNA and, in a proportion of events, the modified allele will be transferred to the chromosome. Early examples of this strategy have been the targeting of mutations generated in the *polA*, *recA* and *hsd* genes to the chromosome. This reaction is very similar to the 'ends-in replacement' described in Fig. 7.13 with the difference that the cloned DNA is not cleaved *in vitro* prior to the first recombination reaction.

recombination when DNA is introduced into cells. Specific strategies have therefore had to be developed to overcome this problem. One strategy that has been successful is to use PCR (polymerase chain reaction) to detect the rare homologous recombinants in a pool of transformed cells and then to enrich for their presence by dividing the pool, growing up and assaying each part by PCR. This **sib-selection** is repeated several times until the

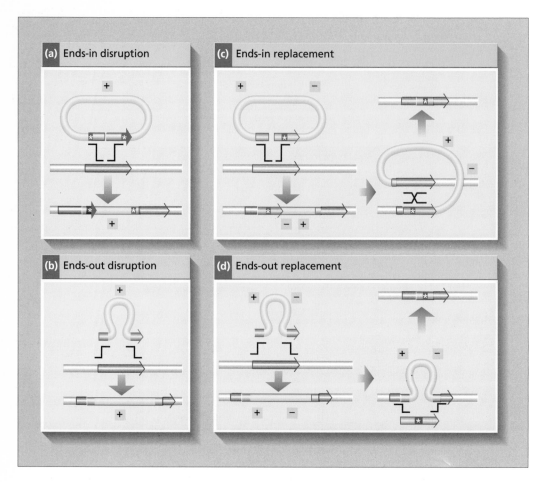

Fig. 7.13 The four basic strategies used in gene-targeting. Targeting is used for two purposes, gene disruption and gene replacement. Gene disruption (a) and (b) is the generation of defective genes via the insertion of cloned DNA. Gene replacement (c) and (d) is the more subtle modification of the gene of interest and requires a two-step strategy. Both disruption and replacement can be executed by 'ends in' or 'ends out' strategies. Cloned DNA is shown in red and the chromosome is shown in grey. The gene of interest is represented by an arrow and parts of it that have been modified or mutated are in full colour and marked with an asterisk. Markers associated with the cloned DNA that allow selection for, or against, their presence under particular selective conditions are represented by + and –, respectively. (a) **'Ends-in' disruption.** Cloned DNA is cleaved once within the homologous sequence to generate two recombinogenic ends and a recombinant is selected by virtue of a marker within the vector. The gene itself must carry mutations in its 3′ and 5′ ends so that two mutant genes are produced following recombination. This strategy suffers from two drawbacks. Firstly, both mutations must inactivate the gene. This is relatively straightforward for a 3′ deletion but 5′ mutations (even deletions) do not always cause inactivation. The second, is that reversal of the disruption can occur by recombination between the directly repeated copies of the gene. (b) **'Ends-out' disruption.** A cloned sequence is firstly modified by insertion of a selectable marker within the gene of interest. A fragment, including the selectable marker and flanking gene sequences, is then generated by cleavage. This DNA is then allowed to recombine with the chromosome, generating a

recombinants are no longer rare. Another strategy is to use an 'ends-out' construct with a **promoterless selectable marker** so that it will only become active when transcription is driven from the target gene. A third approach is to incorporate a gene in the targeting vector which can be selected against and to arrange that illegitimate recombinants inherit this gene but homologous recombinants do not as shown in Fig. 7.14. This has been called **positive-negative** selection. All three of these strategies are really enrichments for the targeted recombinants and leave the investigator with 10–100 clones to screen instead of 1000–10 000.

Mouse embryonic stem cells (ES cells) are now extensively used for gene-targeting because of their pluripotency. These cells can be cultured on a synthetic medium which can be used to select for successfully targeted recombinants. They can then be re-introduced into early mouse embryos where the cells multiply and contribute to all cell types. The result is, therefore, a mosaic individual with both recombinant and parental genotypes. If the ES cells have contributed to the mosaic mouse's germ-line, some of its offspring will carry the targeted gene in all their cells. A new **transgenic** mouse with a desired alteration to its genome will have been constructed. This is illustrated in Fig. 7.15. This technique is beginning to provide new information on the effects of many important genes including those implicated in development, cancer and genetic disease.

There are now many examples of the success of this approach (illustrated in Fig. 7.16 by the studies of the p53 protein which plays a central role in carcinogenesis). In this example, the beauty of using gene-targeting and germ-line transmission of targeted genes is two fold. Firstly, the mutation (a gene disruption) is specific to the targeted locus and no other changes to the genome are made which greatly enhances the unambiguous nature of the results. And secondly, all cell types in the transgenic mouse carry the gene disruption. This means that a specific effect in one tissue or cell type can be investigated in its normal context within the whole animal.

Fig. 7.13 (*Continued*) disrupted gene. (c) **'Ends-in' replacement.** A mutation of interest is introduced into the cloned gene which is cleaved as in (a). A positive selectable marker within the vector allows selection of the integrated DNA which is then allowed to recombine between the directly repeated copies of the gene. Another marker is often included in the vector which allows selection to be applied against the presence of vector sequences and the isolation of the desired product. Because the second stage recombination will occur at several different positions along the gene, the mutation will not always be present in the product. A screen for the presence of the desired mutation is therefore required as a final step. (d) **'Ends-out replacement'.** This is also a two-step construction. Firstly, the gene of interest is disrupted using a DNA fragment containing both a positive and a negative selective marker by selection for the positive marker as in (b). This DNA is then replaced by recombination with a fragment carrying the mutation of interest. In this second step, selection against the negative marker is applied, to allow the isolation of the desired product.

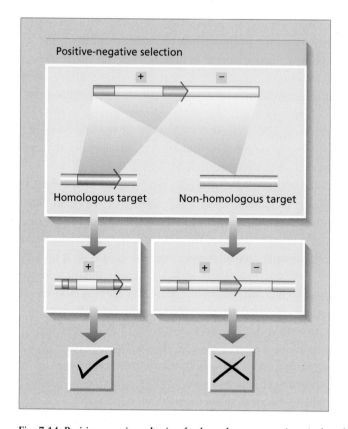

Fig. 7.14 Positive-negative selection for homologous targeting. A cloned gene (denoted by a red arrow) is disrupted by a positive selectable marker (+) in a vector containing a gene which can be selected against (−). A fragment containing the disrupted gene and the gene to be selected against are then introduced into cells and allowed to recombine with the chromosome. If a homologous recombination event occurs, the (−) gene is lost, whereas most illegitimate (end-joining) events will incorporate the whole fragment. If selection is then applied against the (−) gene, homologous recombinants will be significantly enriched. Occasionally, degradation of the DNA will occur prior to illegitimate recombination, allowing a few clones that have not targeted the correct gene to escape the selection.

7.7   Gene-therapy

Gene-therapy is the alteration of an individual's genetic material for therapeutic purposes. It is a new development with significant potential for the treatment of conditions that are currently very difficult or impossible to treat. On the other hand, it raises ethical issues that we must confront. In this context, it is important to understand the difference between the soma and the germ-line (see Fig. 1.4) and to distinguish somatic from germ-line therapy.

Somatic gene-therapy is currently undergoing clinical trials. It is the treatment of an individual's somatic tissue with exogenous DNA and the

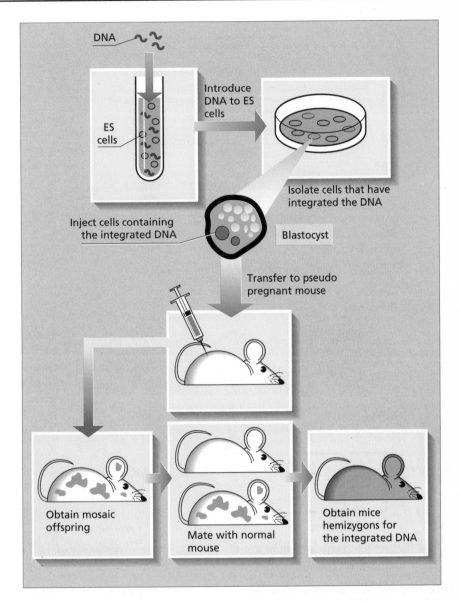

Fig. 7.15 The making of a transgenic mouse using targeted ES cells. DNA is introduced to embryonic stem cells (ES cells) and the required recombinant is selected as described above. The correctly targeted cells are then introduced into a blastocyst and this is transferred to a pseudo-pregnant mouse. The offspring of this mouse will be mosaics, carrying clones of targeted tissue. If the targeted cells constitute the germ line, the targeted gene can be passed on to the offspring of the mosaic mouse. If this happens, the mouse produced will be hemizygous for the integrated DNA (heterozygous for the targeted gene). From such a hemizygous mouse, homozygous progeny can be obtained by selfing.

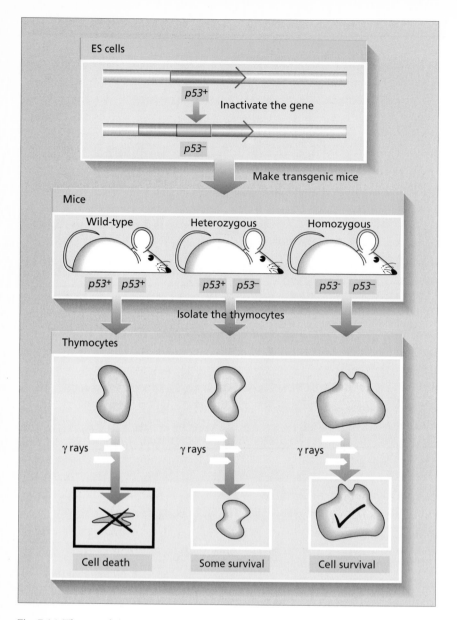

Fig. 7.16 The use of gene-targeting to study the role of p53 in carcinogenesis. (After Clarke *et al.*, 1993 and Lowe *et al.*, 1993.) *p53* is the most commonly mutated tumour-suppressing gene found in human cancers. The p53 protein, which is a transcriptional activator, must play a role in the prevention of tumour formation, but what role? In order to determine the function of p53, it has been necessary to study the physiology of cells lacking the *p53* gene. Gene-targeting has therefore been used to construct a *p53* disruption in mouse ES cells which have been used to regenerate *p53⁻* heterozygous and homozygous mice. Given earlier experiments with cultured cells that had suggested p53 might be involved in programmed cell death, it was a surprise to find these transgenic mice were viable, fertile and immunologically competent. However, further investigation revealed that the thymocytes of these mice were highly resistant to ionizing radiation under conditions where normal *p53⁺/p53⁺* thymocytes underwent programmed cell death. These experiments demonstrate that the p53 protein accelerates cell death in response to ionizing radiation, a property that is likely to explain its tumour-suppressing effect. *p53⁺/p53⁻* heterozygous thymocytes are slightly more resistant to ionizing radiation than wild-type cells which could lead to increased mutagenesis in these cells and a greater propensity to progress to tumour formation by the accumulation of further mutations.

consequences of treatment are limited to the individual undergoing treatment. The issues raised by somatic gene therapy are therefore those familiar to all areas of medicine; the balance between treatment and side-effects, benefit and cost, profit and loss. We can define four strategies for somatic gene therapy, two of which have a primary role for genetic recombination.

The first is the use of disabled virus vectors that are primarily designed to replicate as extrachromosomal elements. These include derivatives of adenovirus and herpes viruses that are non-pathogenic but can maintain cloned DNA sequences in cells over a defined period. Treatment requires the periodic infection of the individual with the genetically modified vector and carries with it the side-effect of an immune reaction to the virus. Recombination is not a primary requirement for this type of treatment but may occur by chance in some cells.

The second is the delivery of DNA directly to cells in the body. This can be either via liposome-coated DNA or with a DNA particle gun. The liposome method relies on fusion of liposomes with the cell membrane to release their contents into the cell and the particle gun involves coating tiny beads of gold with DNA and accelerating them towards a tissue like bullets. The introduction of DNA into cells allows transient expression while the DNA is present and requires repeated treatment to maintain therapeutic levels of gene expression. As with the first method of treatment, recombination is not a primary requirement but may occur in some cells either by illegitimate or homologous mechanisms.

The third is the isolation of a tissue, its treatment outside the body and its subsequent re-introduction. Bone marrow can be used in this way. Trials are underway to treat bone marrow using retroviral vectors (see Section 7.5) carrying important genes and to replace the treated material. Since the integration of retroviral DNA occurs at many locations on chromosomes, there is no control over the sites of insertion. However, once inserted, the DNA should remain stable. Because bone-marrow stem cells are very rare, this method does not provide a permanent cure. As cells containing the inserted DNA become depleted, the treatment must be repeated.

The fourth is the isolation, modification, expansion and re-introduction of tissue-specific stem cells. These experiments are still at the stage of investigation in animal model systems. Here, cells such as primary myocytes (muscle stem cells) or primary keratinocytes (skin stem cells) are treated *in vitro* and cells that have been modified, expanded clonally prior to re-introduction to the body. This method has the advantage of offering a permanent treatment. The modified stem cells will remain alive throughout the individual's life passing on the introduced DNA to all their more differentiated cell types. Furthermore, homologous targeting should be possible using this technique which will ensure expression from the correct chromosomal location.

These approaches are developing very rapidly and the technology is likely to change significantly before methods of choice become clear. Recombination of introduced DNA into the chromosomes has the advantage of producing stable products that promise a more permanent form of treatment than transient expression from extrachromosomal DNA. But expression of integrated DNA has often been found to be very weak, or even closed down. Homologous targeting may overcome this problem but is likely to be most useful for correction of a genetic condition where the appropriate stem cells are available. For expression of a wider range of genes, experiments are underway to determine whether site-specific recombination can be used to target DNA to a known location of good expression that has been manipulated to contain the appropriate recombination target-site.

Germ-line gene therapy is a theoretical possibility but raises considerable ethical issues. It relies on the development of human embryonic stem cell technology which itself would provide the possibility of generating genetically identical individuals. Furthermore, any genetic modifications that were made would be inherited by future generations. In any case, it is arguable that germ-line gene therapy has little positive to offer over pre-natal diagnosis for prevention and somatic gene therapy for treatment. The social, political and psychological consequences of this level of interference with our genetic inheritance must be considered. We have choices that can be made not to use or not to develop certain technologies and we must continue to encourage the debate of the issues so that informed decisions can be made.

7.8   Conclusion

All aspects of recombination from homologous to illegitimate have experimental and potential therapeutic applications. Classically, mapping and strain construction have been the primary applications, but now mutagenesis, targeting and gene therapy are equally important areas where recombination plays a central role. The application of recombination does not necessarily require an understanding of the mechanisms of the underlying reactions but the development of these methods is greatly facilitated by such knowledge. Furthermore, there are always unforeseen pitfalls in applying methods with little understanding of their basis.

Further reading

Berg C. M., Berg D. E. & Groisman E. A. (1989). Transposable elements and the genetic engineering of bacteria. In Berg D. E. & Howe M. M (eds) *Mobile DNA*, pp. 879–926. American Society for Microbiology, Washington.

Brock D. J. H. (1993). *Molecular Genetics for the Clinician*. Cambridge University Press, Cambridge.

Haldane J. B. S. (1919). The combination of linkage values, and the calculation of distances between the loci of linked factors. *J. Genet.* **8**, 299–309.

Kaiser K., Sentry J. W. & Finnegan D. J. (1995). Eukaryotic transposable elements as tools to study gene structure and function. In Sherratt D. (ed.) *Mobile Genetic Elements: Frontiers in Molecular Biology*, pp. 69–100. Oxford University Press, Oxford.

Ott J. (1991). *Analysis of Human Genetic Linkage*, 2nd edn. Johns Hopkins University Press, Baltimore.

Sedivy J. M. & Joyner A. L. (1992). *Gene Targeting*. Freeman, New York.

Soriano P., Gridley T. & Jaenisch R. (1989). Retroviral tagging in mammalian development and genetics. In Berg D. E. & Howe M. M (eds) *Mobile DNA*, pp. 927–938. American Society for Microbiology, Washington.

Stahl F. W. (1979). *Genetic Recombination: Thinking About it in Phage and Fungi*. Freeman, USA.

Suzuki D. T., Griffiths A. J. F., Miller J. H. & Lewontin R. C. (1989). *An Introduction to Genetic Analysis*, 4th edn. Freeman, New York.

References

Bridges C. B. & Olbrycht T. M. (1926). The multiple stock 'Xple' and its use. *Genetics.* **11**, 41–56 (as presented by Strickberger M. W. (1968) *Genetics.* Macmillan, New York.)

Clarke A. R., Purdie C. A. & Harrison D. J. (1993). Thymocyte apoptosis induced by independent pathways. *Nature* **362**, 849–852.

Lowe S. W., Schmitt E. M., Smith S. W. Osborne B. A. & Jacks T. (1993). p53 is required for radiation induced apoptosis in mouse thymocytes. *Nature* **362**, 847–849.

Ott J. (1991). *Analysis of Human Genetic Linkage*, 2nd edn. Johns Hopkins University Press, Baltimore.

Perkins D. (1962). Crossing-over and interference in a multiply marked chromosome arm of Neurospora. *Genetics* **47**, 1253–1274.

Index

Page numbers in *italic* refer to figures